The architecture of the well-tempered environment

I thought I heard Buddy Bolden say,
Open up that window, let the foul air get away!
Open up that window, let the foul air out!
That's what I heard him shout.

Traditional

Reyner Banham

The architecture of the
well-tempered environment

The Architectural Press, London

By the same author:

Theory and Design in the First
Machine Age

Guide to Modern Architecture

The New Brutalism

This book was prepared under a grant from the Graham Foundation of Chicago, Illinois

85139 073 0
First published 1969
© Architectural Press 1969
Made and printed in Great Britain by
William Clowes and Sons, Limited
London and Beccles

Acknowledgements

The preparation of this text has depended heavily on the help and advice of a great many people and organisations. My debts to some are obvious . . .
to the Graham Foundation of Chicago, and to its director, John Entenza, for the Fellowship that afforded time to research and to think
to the Architectural Press, London, to the University of Chicago Press and to Maurice English of the University of Chicago Press, for their encouragement and help
to my wife, whose work on the illustrations has benefited the work with more than a set of well-drawn pictures, for her detective zeal has elucidated many mysteries of plan, section and operation.
But other indebtednesses are less obvious, and some of those listed below may not now remember the stray conversational remark that launched me on a fruitful line of research, though others will certainly recall long and persistent correspondence or other relatively massive operations . . .
George Collins and James Marston Fitch of Columbia University; Philip Johnson, Cervin Robinson, Wallace K Harrison
Stanford Anderson, Henry Millon, Buford Pickens
Peter Carter, Wilbert Hasbrouck of the Prairie School Press, Carl W. Condit, Tom Stauffer, Roy W Shields of the Samuel R Lewis organisation, Mr and Mrs Walter Sobel, Mrs Howard Rosenwinkel, Buckminster Fuller, Elmer Dittrich of Amana Refrigeration Inc.
Esther McCoy, J R Davidson, Richard Neutra, David Gebhard, Randel Mackinson, George A Dudley, Jonathan King of Educational Facilities Laboratory
Lord Llewelyn-Davies and John Weeks, George Atkinson, P. P. Loraine and staff at Claritude/Saint-Gobain, Sir Philip

Magnus, Ralph Hopkinson, James Longmore, Peter Smith, Hugh Chapman, Messrs Dickson & Davis of the Greater London Council Engineer's Dept, and Messrs Blyth and Sutherland of the Architect's Department.
Sidney Cusdin and S. H. Dickinson of Cusdin, Burden and Howitt, Shane Belford, Mr Kelly of the Royal Victoria Hospital, Charles A Brett of Davidson and Co.
Quentin Hughes, Francis M Jones, H. C. Morton
the following libraries and their staffs: New York Public Library; Royal Institute of British Architects, London; the Burnham Library, Art Institute of Chicago; Architecture School, University of Liverpool; the Avery Library, Columbia University; Bartlett School, and Courtauld Institute, University of London.
and Carl Rosenberg, who appeared most opportunely to give help with the manuscript and footnotes.

Contents

1. Unwarranted apology   11

2. Environmental management   18

3. A dark satanic century   29

4. The kit of parts: heat and light   45

5. The environments of large buildings   71

6. The well-tempered home   93

7. The environment of the machine aesthetic   122

8. Machines à habiter   143

9. Towards full control   171

10. Concealed power   195

11. Exposed power   234

12. A range of methods   265

   Readings in environmental technology   290

   Photo credits   291

   Index   292

Illustrations

Larkin Building, Buffalo, by Frank Lloyd Wright 12
Richards Memorial Laboratories, Philadelphia, by Louis Kahn 12
Environmental behaviour of a tent 18
Environmental conditions around a camp fire 20
Professor Jacob's mixing valve for radiators 33
Teale fireplace 33
The Octagon, Liverpool, by Dr John Hayward 36, 37
Air-circulation in room heated by open fire 48
Air movement, temperature and humidity in a lecture room 49
Currents of air in a 'model' hospital ward 49
Heating and ventilating with thermal siphon and with fan extracts 50
Mixing valve for domestic hot air 50
Installation to provide warm air to a store in Boston 51
Air-heating installation, Menominee, 53
Increasing domestic consumption of light 62
Rate of installation of domestic electric lighting 63
Stokesay Court, by Thomas Harris 66
The van Kannel revolving door 74
Royal Victoria Hospital, Belfast, by Henman and Cooper 76–79,
Glasgow School of Art, by Charles Rennie Mackintosh 84, 85
Larkin building, Buffalo, by Frank Lloyd Wright 87, 88, 89
American Woman's Home, by Catherine Beecher 98, 99
Traditional Norwegian farm-house 100
The Gamble House, Pasadena, by Charles and Henry Greene 102, 103
The Baker House, Wilmette, by Frank Lloyd Wright 106
Charles Ross Cottage, Delavan Lake, by Frank Lloyd Wright 110
Architect's own house, Darmstadt, by Peter Behrens 113
Mrs T. H. Gale House, Oak Park, by Frank Lloyd Wright 114, 115
The Frederick C. Robie House, Chicago, by Frank Lloyd Wright 116,
    118, 119
Glass Industry Pavilion, Cologne Exhibition, by Bruno Taut 131
Architect's own house, Berlin, by Bruno Taut 133
Employment exchange, Dessau, by Walter Gropius 135
Light fitting for Dr Hartog, by G. T. Rietveld 136
Architect's own office, Weimar, by Walter Gropius 137

Radio-cabinet and table lamp, by G. T. Rietveld   138

Bioscoop Vreeburg cinema, Utrecht, by G. T. Rietveld   139

Row-houses, Utrecht, by G. T. Rietveld   140

Pantomime, Bauhaus Theatre   141

Light-play, Bauhaus Theatre   142

Victoria Regia water-lily house, Chatsworth, by Sir Joseph Paxton   144

Villa Cook, Boulogne-sur-Seine, by Le Corbusier and Pierre Jeanneret
148, 149

Villa Savoye, Poissy, by Le Corbusier and Pierre Jeanneret   151

Pavillon Suisse, Paris, by Le Corbusier and Pierre Jeanneret   153, 154,
155

Cité de Refuge, Paris, by Le Corbusier and Pierre Jeanneret   156, 157

Test-chamber for *mur neutralisant*, Saint Gobain   161

The Dalsace house, Paris, by Pierre Chareau and Bernard Bijvoet   164–
167

Aluminaire House, Long Island, by Kocher and Frey   168, 169

Milam Building, San Antonio, by George Willis   178, 179

McQuay air-conditioners   186

Lafayette Park apartments, Detroit, by Mies van der Rohe   188, 189

Kips Bay apartments, New York, by I. M. Pei Associates   191

Row-houses, Chicago, by Harry Weese   192, 193

Kuhn and Loeb Bank, New York, by A. M. Feldman   196

Johnson Wax Company Offices, Racine, by Frank Lloyd Wright   197

Arizona Biltmore Hotel, Phoenix, by Albert Chase McArthur   198,
199

Lighting in piano-showroom, Berlin, and lighting for shop-fronts, Los
Angeles, by J. R. Davidson   200

Rudolf Mosse offices, Berlin, by Eric Mendelsohn   201

Grosse Schauspielhaus, Berlin, by Hans Poelzig   201

Herpich Store, Berlin, by Eric Mendelsohn   202

Universum Cinema, Berlin, by Eric Mendelsohn   203

Lovell beach-house, Newport Beach, by Rudolph Schindler   205, 206

Sardi's Restaurant, Hollywood, by Rudolph Schindler   207

Lovell House, Los Angeles, by Richard Neutra   208

Philadelphia Savings Fund Society Building, Philadelphia, by Howe and
Lescaze   210, 212

Harris Trust and Savings Bank Building, Chicago, by Skidmore Owings
and Merrill   211

Corrugated steel floor/ceiling system, by the Mellon Research Institute
214

Perforated Acousti-vent ceiling system, by Burgess Laboratories   215
De Laveaga Elementary School, Santa Cruz, by Leefe and Ehrenkrantz
   217
Skyline Louverall suspended ceiling system   218
General Motors Technical Centre, Warren, by Saarinen, Swanson and
   Saarinen   220, 221
Drake University Laboratories, by Saarinen, Swanson and Saarinen   222
Universal Pictures Building, New York, by Kahn and Jacobs   223
United Nations Headquarters, New York   224, 225
Lever House, New York, by Skidmore Owings and Merril   226, 227
Continental Centre, Chicago, by C. F. Murphy and Associates   229
Architect's own house, New Canaan, by Philip Johnson   230–233
Office interior, London, by Alison and Peter Smithson   234
UN Building, Chamber of the Trusteeship Council, by Finn Juhl   235
UN Building, Foyer of the General Assembly, by Wallace K. Harrison
   236
Unite d'Habitation, Marseilles, by Le Corbusier   238
Olivetti factory, Merlo, by Marco Zanuso   240, 241, 243
La Rinascente Store, Rome, by Albini and Helg   245, 246, 247
Richards Laboratories, Philadelphia, by Louis Kahn   248, 250, 251
Agronomy Laboratories, Cornell University, by Ulrich Frantzen   253
Pharmaceutical plant, Debreczen, by Gulyas and Szendroi   254
Sheffield University extensions, by Alison and Peter Smithson   254
Furniture Industry Headquarters, by Mike Webb   256
Queen Elizabeth Hall, London, by London County Council   258–263
Fremont Street, Las Vegas   271
Portable Theatre, by Victor Lundy and Walter Bird   272, 273
Space capsule, life support system   278
Space suits   279
St George's School, Wallasey, by Emslie Morgan   280, 282, 283
Terrace housing, Sydney   287

1. Unwarranted apology

In a world more humanely disposed, and more conscious of where the prime human responsibilities of architects lie, the chapters that follow would need no apology, and probably would never need to be written. It would have been apparent long ago that the art and business of creating buildings is not divisible into two intellectually separate parts—*structures*, on the one hand, and on the other *mechanical services*. Even if industrial habit and contract law appear to impose such a division, it remains false.

If there is any division at all that can be tolerated in a humane consideration of architecture, it might be between those parts of structure that combine with certain mechanical services to provide the basic life support that makes a viable or valuable environment, and those parts of structure that combine with certain other mechanical services to facilitate circulation and communication—of persons, information and products.

The fact that the outpourings of a radio may be understood as information or environmental background, that the flow of hot water through a pipe may be seen as contributing to the maintenance of an environmental condition or the transmission of a useful product, should warn us that the making of even the division proposed above is open to serious questioning, though the validity of this division for the purposes of the present book, which discusses the architecture of environment, should emerge as the argument proceeds.

Yet architectural history as it has been written up till the present time has seen no reason to apologise or explain away a division that makes no sense in terms of the way buildings are used and paid for by the human race, a division into structure, which is held to be valuable and discussible, and mechanical servicing, which has been almost entirely excluded from historical discussion to date. Yet however obvious it may appear, on the slightest reflection, that the history of architecture should cover the whole of the technological art of creating habitable environments, the fact remains that the history of architecture found in the books currently available still deals almost exclusively with the external forms of habitable volumes as revealed by the structures that enclose them.

The main topic of the present study has therefore only impinged upon the attention of architectural historians when it has incontrovertibly affected the external appearance of buildings, the most notable case being that of the Richards Memorial Laboratories in Philadelphia, by Louis Kahn. By giving monumental external bulk to the accommodations for mechanical services, Kahn forced architectural writers to attend to this topic in a way that no recent innovation in the history of servicing had done. No matter how profound the alterations wrought in architecture by the electric lamp, or the suspended ceiling (to cite two major instances of revolutionary inventions), the fact that these alterations were not visible in outward form has denied them, so far, a place in the history of architecture.

Above: Larkin Administration Building, Buffalo, N.Y. 1906, by Frank Lloyd Wright.
Below: Richards Memorial Laboratories, Philadelphia, Pa., 1961, by Louis Kahn.

Yet what was visibly manifest in the Richards Laboratories, had been equally visible and manifest in Frank Lloyd Wright's Larkin Building in Buffalo, more than half a century earlier. Few architectural writers have made anything of those strong and monumental forms that Wright gave to the external expression of his pioneer system of mechanical servicing, however, except to cite them as the purely formal source of the external service-works of the Richards Laboratories.

So shallow an interest in so profound a building was both inevitable and predictable however; the art of writing and expounding the history of architecture has been allowed—by default and academic inertia—to become narrowed to the point where almost its only interest outside the derivation of styles is haggling over the primacy of inventions in the field of structures. Of these two alternatives, the study of stylistic derivations now predominates to such an extent that the great bulk of so called historical research is little more than medieval disputation on the number of influences that can balance upon the point of a pinnacle.

As a result, a vast range of historical topics extremely relevant to the development of architecture is neither taught nor mentioned in many schools of architecture and departments of architectural history. Some are external to the buildings—patronage, legislation, professional organisation, etc.—others are internal—changes in use, changes in users' expectations, changes in the methods of servicing the users' needs. Of these last, the mechanical environmental controls are the most obviously and spectacularly important, both as a manifestation of changed expectations and as an irrevocable modification to the ancient primacy of structure, yet they are the least studied.

Thus, when the research for the present study was first put in hand, the intention was to write a purely architectural history; to consider what architects had taken to be the proper use and exploitation of mechanical environmental controls, and to show how this had manifested itself in the design of their buildings. To achieve this, some grounding in the purely technical history of these controls was obviously required, but I discovered that no comprehensive study of the topic could be found. The one work that was persistently recommended to me as having covered the ground or exhausted the topic, was Sigfried Giedion's *Mechanisation Takes Command*[1] of 1950. It proved, however, in no way to deserve such a reputation—a point to which this argument must return.

[1] London and Cambridge, Mass., 1950.

What needs to be said here and now is that although there can be no doubt that my view of the topic has been vastly enriched by my enforced studies of primary source material (trade catalogues, lectures to professional societies, specialist periodicals, etc.) the absence of any general and compendious body of study in the field leaves little chance of estimating how balanced and comprehensive is the view that I have derived from these readings.

The matter probably cuts deeper than this, because the absence of a body of studies means that the architectural, as well as the technical, aspects may be off balance. Thus, the average producer of a pinnacle-point type of doctoral dissertation on some such subject as 'The Influence of the Drawing Style of Mart Stam on the Aesthetics of *Elementäre Gestaltung*' is a scholastically secure man. He may be setting out to make drastic modifications in the balance of reputations of a group of architects working in a certain place at a certain time, but a known balance of reputations already exists for him to modify, because a continuing body of academic work keeps that balance under review in lecture, seminar and learned paper.

But step outside the security of that continuing body of work, and not only is there no balance of reputations, there are no reputations at all. Nobody knows who were the true masters and innovators, or who merely rode the coat-tails of genius. Ask a historian of modern architecture who invented the *piloti*, and he can tell you. Ask him who invented the (equally consequential) revolving door, and he cannot. Ask who were Baron von Welsbach, Samuel Cleland Davidson or what was first done on the façade of the *West-End* Cinema in Leicester Square, London, and the answer will come very haltingly if at all, and yet these are all matters deserving more than a footnote in any history of what has really happened in the rise of modern architecture.

In such conditions of ignorance and insecurity, and the sheer paucity and poverty of academic discourse on the topic, the reputation of *Mechanisation Takes Command* is perhaps understandable. Even James Marston Fitch, whose sagacious observations on environment and technology have been a constant inspiration to my studies, speaks of Giedion's book as 'a new and revealing study of American technology' despite the fact that his own published works constantly reveal the shallow and unconsidered nature of Giedion's observations.

The true fault of the book lay in its reception. Awed by the immense reputation of its author, the world of architecture received *Mechanisation Takes Command* as an authoritative and conclusive statement, not as a tentative beginning on a field of study that opened almost infinite opportunities for further research. In the ensuing twenty-odd years since its publication, it has been neither glossed, criticized, annotated, extended nor demolished. 'Giedion,' one is told 'hasn't left much to be said.'

This present book represents a tiny fraction of what Giedion left unsaid. This too is a tentative beginning, whose shortcomings, I have no doubt, will become manifest as research proceeds, especially since it suffers from at least one defect in common with Giedion's—the use of the concept of 'the typical.'

The chapters that follow are not exhaustive, therefore they are not definitive. In the light of partial knowledge one cannot specify with certainty, only typify with hope. That is, all one can really do is to indicate the sort of work that was done in a particular period of time, and select a particular building that seems to typify the kind of architecture done with that technique at that time. But in the absence of encyclopaedic knowledge or a going body of research and discussion, it is extremely difficult to be confident that one has picked the most typical building, or the best of a number of buildings exemplifying the same point. Matters of exact primacy in date, who thought of what first, are even harder to fix under these circumstances, but on this point, and in the context of this study, the use of the typical rather than the exactly definitive, can be defended.

While Patent-Office records, of the sort exploited by Giedion and his students in compiling *Mechanisation Takes Command*, make legal primacy of invention capable of being fixed with documentary certainty, such exact dates may be totally valueless in studying the history of architecture. In the practical arts like building, it is not the original brainwave that matters as much as the availability of workable hardware, capable of being ordered ex-catalogue, delivered to the site and installed in the structure. Thus the early patents for fluorescent lighting are almost inconsequential for the history of architecture, but the commercial availability of reliable tubes some thirty-six years later was to be of the utmost consequence. More confusingly, it is possible that one or two major buildings were being air-conditioned (in some senses) two or three years before the earliest air-conditioning patents, and before the phrase 'air-conditioning' had even been coined.[2]

[2] see chapter 9.

In conditions such as these, it may be unwise at present to try to establish absolute primacy of installation or exploitation, and pointless to lavish too much attention on primacy of invention. It has seemed better, in many cases, to settle for a building which appears to sum up forward thinking and progressive practice and let it stand as typical of the best or most interesting work being done at the time, but not to attribute to the concept of typicality those overtones of a platonic absolute implicit (and explicit) in Giedion's elevation of Linus Yale to the status of the very *type* of the Yankee inventor. The use of typicality in the chapters that follow is purely illustrative, the buildings singled out for mention tend less to be the first of their class, than 'among the first.'

This too seems just; this is less a book about *firsts* than about *mosts*. The invention and application of technological devices is not a static and ideal world of intellectual discourse; it is (or has been) impelled forward by the competitive interaction of under-achievers and over-achievers—who might even be one and the same person, for some breakthroughs in application were achieved without matching breakthroughs in invention. But nothing would have been broken through without some extremism of method, and extravagance of personality.

Le Corbusier might admonish in 1925 that 'an engineer should stay fixed and remain a calculator, for his particular justification is to remain within the confines of pure reason . . .'[3] but the fact remains that many of Le Corbusier's own buildings would have been unbuildable or uninhabitable had engineers ever heeded his advice, instead of pursuing their own eccentric and monomaniac

[3] *Urbanisme*, Paris, 1926, p 48.

goals without regard for professional demarcations and social conventions. The history of the mechanisation of environmental management is a history of extremists, otherwise most of it would never have happened. The fact that many of these extremists were not registered, or otherwise recognised as architects, in no way alters the magnitude of the contribution they have made to the architecture of our time. Perhaps finding such men a proper place in the *history* of architecture will be some help in resolving the vexed problems of finding their proper place in the *practice* of architecture.

2. Environmental management

The surviving archaeological evidence appears to suggest that mankind can exist, unassisted, on practically all those parts of the earth that are at present inhabited, except for the most arid and the most cold. The operative word is 'exist'; a naked man armed only with hands, teeth, legs and native cunning appears to be a viable organism everywhere on land, except in snowfields and deserts. But only just; in order to flourish, rather than merely survive, mankind needs more ease and leisure than a barefisted, and barebacked, single-handed struggle to exist could permit.

A large part of that ease and leisure comes from the deployment of technical resources and social organisations, in order to control the immediate environment: to produce dryness in rainstorms, heat in winter, chill in summer, to enjoy acoustic and visual privacy, to have convenient surfaces on which to arrange one's belongings and sociable activities. For all but the last dozen decades or so, mankind has only disposed of one convincing method for achieving these environmental improvements; to erect massive and apparently permanent structures.

Partial solutions to these problems have always been offered by alternative methods such as wearing a coat in the rain; getting in a tent out of the sun, or gathering around a camp-fire in the cool of evening. But a coat is an unsociable solution, a tent is short on acoustic privacy even though it may be adequate to keep off prying eyes, and a camp fire, while it can provide heat and light enough to make a useful area of ground habitable, is short on all sorts of privacy and offers no protection against rain.

But, over and above considerations of this kind, one must observe a fundamental difference between environmental aids of the structural type (including clothes) and those of which the camp-

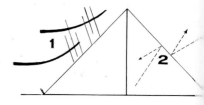

Environmental behaviour of a tent.

1. Tent membrane deflects wind and excludes rain
2. Reflects most radiation, retaining internal heat, excluding solar heat, maintaining privacy

fire is the archetype. Let the difference be expressed in a form of parable, in which a savage tribe (of the sort that exists only in parables) arrives at an evening camp-site and finds it well supplied with fallen timber. Two basic methods of exploiting the environmental potential of that timber exist: either it may be used to construct a wind-break or rain-shed—the structural solution—or it may be used to build a fire—the power-operated solution. An ideal tribe[1] of noble rationalists would consider the amount of wood available, make an estimate of the probable weather for the night—wet, windy, or cold—and dispose of its timber resources accordingly. A real tribe, being the inheritors of ancestral cultural predispositions would do nothing of the sort, of course, and would either make fire or build a shelter according to prescribed custom—and that, as will emerge from this study, is what Western, civilised nations still do, in most cases.

The acquisition of such predisposing cultural habits depends, obviously, on the previous experience of the tribe or civilisation, and this experience could have been painful. In terms of capital expenditure, a structural solution will usually involve a large, and probably hurtful, single investment, while the power-operated solution may represent a steady and possibly debilitating drain on resources that are difficult to replenish. Most 'pre-technological' societies have little choice in this matter, since they are usually short of combustibles or other sources of usable power. For this reason, all the major civilisations to date, those that have shaped world architecture, have demonstrably, and demonstratively, relied on the construction of massive buildings to fulfil their environmental needs, both physical and psychological.

The consequence is that architects, critics, historians and everyone else concerned with environmental management in civilised countries, lack a range of spatial experience and cultural responses that nomad people have always enjoyed. Cultures whose members organise their environment by means of massive structures tend to visualise space as they have lived in it, that is bounded

[1] This tribe has respectable ancestors, who may be found in Laugier's *Essai* and Le Corbusier's *Vers Une Architecture*. The transformations of the basic parable in which they appear, from the Age of Reason to the present text, may afford the interested reader some insights into our changing conception of the technical and social nature of architecture.

and contained, limited by walls, floors and ceilings.[2] There are, obviously, reservations and quibbles that can be raised against this proposition, but its general truth may be observed in many things, such as the persistent manner in which architects and designers visualise 'free' or 'unlimited' space as retaining the rectangular format of walled rooms—Frederick Kiesler's *Cité dans l'Espace* of 1924 is an obvious instance.

Against this, societies who do not build substantial structures tend to group their activities around some central focus—a water hole, a shade tree, a fire, a great teacher—and inhabit a space whose external boundaries are vague, adjustable according to functional need, and rarely regular. The output of heat and light from a camp-fire is effectively zoned in concentric rings, brightest and hottest close to the fire, coolest and darkest away from it, so that sleeping is an outer-ring activity, and pursuits requiring vision belong to the inner rings. But at the same time, the distribution of heat is biased by the wind, and the trail of smoke renders the downwind side of the fire unappetising, so that the concentric zoning is interrupted by other considerations of comfort or need.

Without pursuing the consequences of these experiences, which may prove to be of fundamental relevance to power-operated environments, further than the exiguous anthropological information warrants, one can still observe that they are experiences that do not enter into the traditions of architecture, even those of modern architecture which is largely concerned with power-operated environments. The traditions of architecture, as we commonly understand the concept, have been forged in societies and cultures that are committed to massively structural methods of environmental management. Furthermore, the accumulation of capital goods and equipment needed to produce even a moderate level of civilised culture in pre-technological societies, required that building materials be treated as if valuable and permanent. It was necessary not only to create habitable environments, but to conserve them. There was rarely any shortage of physically or

[2] see the observations of Paul Scheerbart in chapter 7.

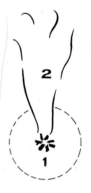

Environmental conditions around a camp fire.

1. Zone of radiant heat and light
2. Downwind trail of warmed air and smoke

culturally necessary functions queueing up for the available stock of roofed spaces. Buildings were made to last, and had to be, in order to produce a sufficient return in terms of shelter performance over the years to justify the expenditure of labour and materials that went into them.

Architecture came to be seen as the conscious art of creating these massive and perdurable structures, and came to see itself professionally as no more than that art, which is one of the reasons for its present problems and uncertainties. Societies—through whatever organs they see fit, such as state patronage or the operation of the market—prescribe the creation of fit environments for human activities; the architectural profession responds, reflexively, by proposing enclosed spaces framed by massive structures, because that is what architects have been taught to do, and what society has been taught to expect of architects.

But such structures may be open to objection on a number of grounds; culturally they may be over-emphatic, economically they may be too expensive, functionally they may be intractable to alteration, environmentally they may be incapable of delivering the performance for which society had hoped. All these objections have grown in force as more technological societies have emerged in the northern hemisphere and sought to establish outposts nearer the equator. But the architectural profession has had little to offer beyond further variations upon massive structure, and has normally responded as if these constituted the unique and unavoidable technique for dealing with environmental problems.

In truth, they never had been the unique and unavoidable technique. A suitable structure may keep a man cool in summer, but no structure will make him warmer in sub-zero temperatures. A suitable structure may defend him from the effects of glaring sunlight, but there is no structure that can help him to see after dark. Even while architectural theory, history, and teaching have proceeded on the apparent assumption that structure is sufficient for necessary environmental management, the human race at large

has always known from experience that unaided structure is inade-
quate. Power has always had to be consumed for some part of every
year, some part of every day. Fires have had to be burned in
winter, lamps lit in the evening, muscle power for fans, water
power for fountains used in the heat of the day.

The design of buildings has always had to make some provision
in plan and section, for these marginal consumptions of environmental power—chimneys for smoke, channels for water. Some
architects, like the Adam brothers, made ingenious use of 'left
spaces' in plan to provide concealed access for servants to light
lamps and candles. In general, however, such provisions were of
little consequence either in outlay or visible bulk; architecture
could continue to treat them as matter for footnotes and appen-
dices (Alberti's generous views on chimneys notwithstanding) and
cleave to the massive structure of walls and roofs as its real
business.

The word 'massive' deserves to be emphasised. In the Mediterranean tradition, from which most Western architecture is
directly descended, the need to render society's shelter-investment
permanent—or, at least, perdurable—was normally answered by
making it massive. Thick and weighty structures are less easily
overthrown by storm or earthquake, less maimed by fire or flood.
But such constructions bring with them environmental advantages
that had become so customary in three millennia of European
civilisation, that they were falsely supposed to be inherent in all
structural techniques, and there were baffled complaints when they
were found to be absent from light-weight methods promoted out
of futuristic enthusiasm for the 'Machine Age.'

The outstanding advantages are acoustic and thermal. A thick
and weighty structure offers better sound-insulation, better thermal insulation and—equally important—better heat storage
capacity. This last quality of massive structure has probably played
a larger part in rendering European architecture habitable than is
commonly acknowledged. The ability of massive structure to

absorb and store heat that is being applied to it, and to return that heat to the environment after the heat source has been extinguished, has served European architecture well in two ways: the mass of masonry in a fireplace, chimney-breast and chimney, has served to store the heat of the fire during the day while the fire burns, and to return it slowly to the house during the chill of the night when the fire has burned out.

Alternatively, the thick walls of a house in a hot climate will hold solar heat during the day, slowing down the rate at which the interior becomes hot, and then, after sunset, the radiation of that heat into the house will help to temper the sudden chill of evening. In more sophisticated forms that use glass as a filter to discriminate between light-energy, which is allowed to pass, and heat energy, whose passage is barred, similar effects of thermal storage are used in the normal green-house, and the whole technique might well be termed the 'Conservative' mode of environmental management, in honour of the 'Conservative Wall' at Chatsworth, devised by that master-environmentalist Sir Joseph Paxton, in 1846.

This conservative mode seems to have become the ingrained norm of European culture, though it has always had to be modified, drastically in humid or tropical climates, less obviously for every-day use, by the 'Selective' mode which employs structure not just to retain desirable environmental conditions, but to admit desirable conditions from outside. Thus a glazed window admits light but not rain, an overhanging roof admits reflected sunlight, but excludes the direct sun, a louvered grille admits ventilating air but excludes visual intrusions.

Traditional construction has always had to mix these two modes, even without recognising their existence, just as it has always had to incorporate the 'Regenerative' mode of applied power, without fully acknowledging its presence. But if these various modes should not be too sharply distinguished in traditional practice, there is an important geographical or climatic consideration that distinguishes solutions that are more conservative from those that are more selective, and an historical watershed that separates both of these from solutions that are primarily regenerative.

The conservative mode suits mainly dry climates, including those that are dry and cold, as well as Mediterranean or semi-desert conditions; the selective mode finds its most needed employment in moist climates, especially in the tropics. Humidity is the crucial factor here, even more than latitude or temperature, as can be seen very clearly in the traditional architecture of the southern United States. In the humid south-east, its attributes have been summed up by James Marston Fitch as

1. Elevated living floors . . . offering maximum exposure to prevailing breezes.
2. Huge, light-mass parasol-type roofs to shed sub-tropic sun and rain.
3. Continuous porches and balconies to protect walls from slanting sun and blowing rain.
4. Large floor-to-ceiling doors and windows for maximum ventilation.
5. Tall ceilings, central halls, ventilated attics for warm-weather comfort.
6. The louvered jalousie, providing any combination of ventilation and privacy . . . etc.[3]

This is a classic characterisation of the selective mode, preoccupied with admitting moving air, and excluding almost every other aspect of the external environment. The conservative mode that prevails in the hot, dry desert south-west has yet to find so masterly a summation,[4] but its crucial differences are immediately apparent —the massive adobe walls and the relatively smaller openings to insulate indoors from out, the carry-over of shaded courtyards from the Spanish Mediterranean tradition. Whatever part differing cultural traditions, as between Louisiana French and Mission Spanish, may have played in this distinction of environmental methods, neither would have survived had it been totally unsuited to the local conditions, and the critical difference in local conditions is humidity.

And of all the factors involved in environmental management, humidity has, for most of architectural history, been the most

[3] in his essay 'The Uses of History', in *Architecture and the Esthetics of Plenty*, New York, 1961, pp 244–245.

[4] Ralph Knowles and his students at the University of Southern California have made a start, however, with their studies of the thermal performance of Indian pueblos such as Mesa Verde.

pestiferous, subtle and elusive of control. While the deficient humidity of an overdried climate can be crudely made good by splashing water about and using shade to reduce evaporative loss, the removal of excess water from the atmosphere has so effectively defied all pre-technological efforts, that it has usually made better sense for those who could afford it to move elsewhere—the British in India retiring to hill-stations like Simla, New York business men with lung complaints to Colorado.

For excess moisture, only a regenerative solution, consuming power, has so far proven effective. Hence the historical, rather than geographical, division between the two main methods of dealing with humid climates. Structural solutions of the Louisiana type discussed above could only be replaced when certain crucial advances in power technology and its control had been achieved.

These advances were part of a general revolution of environmental technology in which humidity control was a late development, and if there is a critical year in that revolution, it is 1882, the year of the domestication of electric power, an achievement that confirmed previous crude environmental advances, and laid the essential foundations for more sophisticated later ones, such as the control of humidity on which air-conditioning depends. It was this revolution that first posed the problem of alternatives to structure as prime controller of environment, and introduced the regenerative mode as a serious rival to the conservative and selective modes, rather than their modest hand-maiden.

It is a fact—though not an easy one to interpret—that the most vital advances into the regenerative mode were made in that area of 'European' architecture that was least devoted to massive construction—North America. This may have depended on the simple coincidence that the abundant timber of which lightweight American houses were built, also provided abundant fuel for the high performance Franklin stoves and Rumford fireplaces that heated them, or it may be that there is a more directly causal connection, and the skimpy thermal performance of these timber buildings made the invention of high-performance, quick-heating stoves environmentally necessary. Or it may have been something even more coincidental than either of these propositions—that these ingenious devices were almost invented for the sake of inventing something or improving an existing device, without any specific reference to the context in which they were to perform.

Whatever happened, it is clear that by the later nineteenth-century, the North Americans had acquired habits and skills in the deployment of regenerative environmental aids that were beginning to add up to an alternative tradition. The importance of this developing regenerative tradition can be seen in the shifting centre of environmental invention as the century proceeded. Coal-gas as a source of domestic environmental power for light and heat is a purely European development, its founding fathers being Philippe Lebon in France, F. A. Winzer in Germany and England, William Murdock in England. But at the other end of the nineteenth century, there can be no doubt that Edison was the true father of the electric light, and Carrier of air-conditioning. Many European inventors, of course, contributed key devices to these regenerative aids, but their development into practicable systems is a purely American story in both cases.

The history of environmental management by the consumption of power in regenerative installations, rather than by simple reliance on conservative and selective structures, is thus a predominantly American history, at least in its pioneering phases. This is in no way a judgement upon the ingenuity or determination of European architects and inventors; it is more a reflection of the unusual problems and advantages of US conditions. The problems were those of lightweight structures in extreme climates wherever Americans built in wood, and the advantages were those of the relatively lightweight culture that many Americans took westward with them into a zone of abundant power.

Of all these considerations, the lack of the encumbrances of a massive culture (physically or figuratively speaking), may have been the most important. It is striking how often events in the USA are not so far in advance of Europe technically, but the Americans appear to have been more aware of what they were doing, and thus to make a better job of it. To anticipate a comparison to be made in a later chapter, one may cite again that masterpiece of the architecture of the well-tempered environment, the Larkin building. In physical and physiological fact it was less advanced than the Royal Victoria Hospital, Belfast, completed some two years earlier, but the advances achieved at the RVH seem rather accidental, and its quality as architecture is barely to be mentioned in the same breath as the Larkin building's.

Doubtless, Wright's towering genius had a great deal to do with this difference in quality, but that genius fed upon a far greater experience in the handling of regenerative tackle than any of his European contemporaries could boast, within the context of a culture that was far more convinced of the need for their exploitation. Familiarity is the key, without a shadow of doubt. There is nornormally a time-lag—sometimes of decades—between a mechanical device becoming available, and its full-blooded exploitation by architects.

This has less to do, directly, with problems of development in the device itself, than with the need for architects to make themselves acquainted with it. In their role as creators of actual physical environments, architects have to be both cautious and practical. They have to see something in use, sometimes for as much as a generation, before they feel the confidence to extrapolate new and radical uses for it, knowing that their clients will never forgive nor forget if anything goes wrong, even if it is the inexperience or improvidence of the client himself that causes the malfunctioning.

So, technological potential continuously runs ahead of architectural performance. The gap between the two is commonly occupied by environmental experimentation in fields not commonly regarded as architecture—greenhouses, factories, transportation. Almost four decades separate the first industrial uses of air-conditioning from its confident employment in the kind of architecture that is designed by famous architects, but these long intervals involve not only physical experimentation, but much speculation and brainstorming as well, in which a climate of ideas is generated that makes the eventual architectural exploitation of the particular technology become thinkable.

These speculations do not take place in a philosophical or professional vacuum. Commercial and personal interests are deeply involved, axes are ground, factions are served. Thus most of what emerges from the technical side proves to be overt or covert sales-promotion literature, what emerges from the architectural side is often propaganda directed at clients, professional self-criticism or attempts to twist the future development of the art.

Even where a visionary without a professional interest emerges, as in the case of Paul Scheerbart and his book *Glasarchitektur*,[5] the propaganda aim remains clear, the intention to mould the world nearer to heart's desire is manifest. For the environment touches man where it hurts—and it hurt Scheerbart deeply—so that the literature of the subject is very closely entangled indeed with practicalities. Much of that literature is of such quality and interest that it could probably stand being discussed in isolation as a separate branch of architectural writing, but to do so would be to deprive it of its reality. None of the chapters that follow is con-concerned solely with theory, none solely with practice. The words uttered, like the buildings erected, are exchanges in the close dialogue of technology and architecture, a dialogue that has become closer and more involved throughout the period covered by this book the period in which the possibility of a purely re-generative architecture has emerged for the first time in human history.

[5] see chapter 7 again.

3. A dark satanic century

An understanding of the way in which radical improvements in environmental technology came about requires a knowledge, not only of the mechanical opportunities and cultural advantages of the improvers and inventors, but also of the atmosphere in which they worked. The word 'atmosphere' is to be read literally. Whatever complaints may circulate today about air-pollution, as about traffic-congestion, we tend to forget that there is ample evidence that both were conspicuous evils of the nineteenth-century urban scene. Our common mid-twentieth-century habit of blaming both on the automobile, like the nineteenth-century habit of blaming them on the railways, the factory system, or other fashionable evils, ignores the fact that the root causes are simply the crowding of men to-gether into restricted spaces. While it was necessary for men, in Aristotle's phrase, 'to come together in cities in order to lead the good life', those cities, by virtue of the coming together of men, would become places of pollution and congestion. The contribu-tion of the industrialising nineteenth century was to bring even more people together at even higher concentrations, and to mark the gravity of the situation by means of new industrial wastes that gave unavoidable visible and olfactory form to the threat to health.

Phrased in the coolest possible terms, the working and living conditions of men in industrialised societies gave rise to en-vironmental problems of the utmost urgency and baffling novelty. The sheer size and human density of settlements posed problems of waste disposal, and threat of epidemic (a threat tragically often fulfilled) that called for powerful legal action. The accumulation of large numbers of workers and mechanical plant in such places as factories and mines called for more than Factory Acts and simi-lar legislation; sanitary and ventilating techniques had to be renovated and improved by radical inventions. The length of the working day required an unprecedented provision of artificial light (with its attendant fire-risks) even in structures above ground like shops and office-blocks. Furthermore, the pollution of the external atmosphere by the waste products of industry and primitive power-generation, and the matching pollution of the indoor atmosphere by human respiration and the inefficient combustion of illuminants, both served to aggravate problems that would have been almost intolerable without them.

However, the mere fact that the combustion of illuminants was inefficient and that most of the outdoor pollutants were wastes, gave an immediate and compelling motive for environmental improvement without waiting upon humanitarian legislation or political action by the victims of pollution. The inefficiency and waste represented lost profits to somebody, and the prospect of gain to any ingenious inventor who could reduce those losses. Below two dramatic panoramic photographs of Chicago wreathed in impenetrable palls of smoke, with the headline *Wastefulness*, the heating engineer M. C. Huyett declared in 1895

> While looking from a window on the fifteenth floor of the Monadnock building and observing smoking chimneys and escaping steam, the above headline (ie., Wastefulness) was suggested, because in it was expressed the economic condition presented to sight. Crossing the Chicago River and seeing hot water and steam from the sewer pipes of individual buildings emptying into the river, and when walking along the streets and seeing steam escaping from manholes, fixed in mind 'Wastefulness' and suggested the thought; 'What does the needless waste from these sources cost Chicago daily—$50,000–$100,000?'[1]

[1] *Mechanical Heating and Ventilating*, Chicago, 2nd ed., 1895, p 76.

Not only this, but there was obviously a matching wastefulness of human resources. However much, or little, nineteenth-century mill-owners and factory bosses may have regarded child-labour as an expendable commodity, that commodity was little use, even while fresh and unmaimed, if it could not see or breathe. One of Willis Carrier's earliest industrial air-conditioning installations was for the purpose of laying a fog of tobacco dust in a cigar factory, where conditions were so bad that the efficiency of workers was seriously affected. The foulness of the average nineteenth-century industrial environment is now almost beyond twentieth-century belief; its killer smogs and constant soot-fall little more than legends kept alive by the entertainment industry as picturesque effects in Sherlock Holmes stories. Yet the incidence of compulsive hand-washing in the early literature of psychoanalysis suggests that atmospheric pollutants may have corroded the minds, as well as the bodies, of those who had to endure these conditions.

If the elimination of profitless waste was one ever-present incentive to environmental improvement, the mere preservation of human life, and sufficient health for survival, was another, and ultimately more important one. As early as the eighteen-sixties, the difference in health of those working in controlled—even crudely controlled—environments and those in relatively uncontrolled ones, was a matter of public record. Ernest Jacob, in his posthumous *Ventilating and Warming*, cites Sir John Simon's report to the Privy Council

> In the year (1863) the deaths from consumption in country districts being taken as 100, the deaths in Manchester counted 263, and in Leeds 218. The greatest mortality took place among printers and tailors, classes who work largely by night, requiring a strong light, which necessitates the burning of much gas. On the other hand, contemporary statistics showed that the miners of Northumberland and Durham, where the pits were freely ventilated, formed an important exception to this rule . . .[2]

Since the safeguarding of health was so important an incentive to environmental study and reform, there should be no surprise at the important part played by medical men in these fields. What may occasion surprise nowadays is that their progressive activities involved direct action in the field of building. Their writings often reveal an intimate practical knowledge of the environmental performance of buildings, an expressed contempt for the architectural profession's apparent indifference to such matters, proposals for the improvement of building-design, and

[2] *Notes on the Ventilation and Warming . . . etc.* (SPCK Manuals of Health), London, 1894, pp 19ff. Professor Jacob, who taught at the Yorkshire College, Leeds, died shortly before his little book was published.

even the construction of reformed buildings by doctors themselves.

Thus Professor Jacob, who was quoted above, had no doubt at all that, as a pathologist, he was far better informed on such matters as heating and ventilating than were the architects whose work he had to visit, professionally or privately. The views of architects on environmental matters he regarded as little better than super-stitious

> . . . in most cases architects are content to introduce an occasional air-brick or a patent device called a 'ventilator'. . .
> Real ventilation is so uncommon that . . . the architect usually thinks this object has been attained if some of the windows can be opened. Some think that the presence of 'ventilators', especially if they have long names and are secured by 'Her Majesty's letters patent', ensures the required end. We may as well supply a house with water by making a trap-door in the roof to admit rain.[3]

This last point is of some importance in the context of the common state of architecture in the seond half of the nineteenth century. In the effective absence, from most buildings, of any system of ducted and force-fed ventilation (comparable with piped water under a sufficient head of pressure to make it go where it was needed) the movement of air was an almost uncontrollable function of the entire building structure, complete with its ancillary services and external weather conditions—the shade of a single tree, the closing of a door or the lighting of a fire in a spare bedroom might make the difference between tolerable and intolerable conditions. On the effect of innovations in ancillary services Jacob observes, for instance, that in concert halls

> Electric light being generally used, the heat from (gas operated) sun-burners—which were formerly used for lighting purposes—is not now available for ventilation . . .[4]

and, again, on the uses of external weather as an environmental aid

> A perfectly still day is the time when the greatest change of air is required, and the time when all wind-actuated schemes fail.[5]

The concept of the total involvement of the entire structure, its

[3] op. cit., p 28.

[4] ibid., p 94.

[5] ibid., p 57.

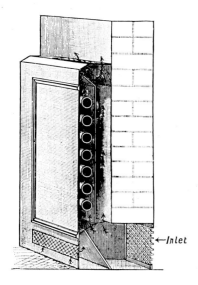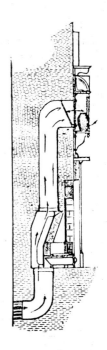

←Inlet

Left: Professor Jacob's 'mixing valve coil-box for radiators'. Centre and right: Teale fireplace with warming chamber and concealed exits for warmed air in overmantel.

inhabitants and their activities, in the processes of ventilation and the distribution of heat, is what Professor Jacob's slim volume was all about. It was written primarily for the guidance of clergy, ecclesiastical building committees and church architects. The environmental insufficiencies of buildings for religious ritual and study always bring out his most caustic and characteristic blend of intellectual scorn and humane sympathy:

> The worst offenders against the laws of health are those responsible for the building of churches and other places of worship. The reason for this is not far to seek . . .[6]

[6] ibid., p 26.

Then follow some admirable examples of 'whole-building' environmental analysis:

> A church is built on a conventional plan, fixed in mediaeval times, when churches were less crowded, services shorter, and above all, at a time

when there was no lighting by gas . . . It is generally built in the form of a nave and side aisles, lighted by clerestorey windows. This gives, including the chancel, four ceilings of different heights, making it most difficult to extract the air at the level of the roof. The clerestorey windows chill the air as it rises, and send it down in the form of a cold douche on the heads of the congregation. The roof is lofty and dark, necessitating a large amount of light, and as a rule about twice as much gas is burned for lighting purposes as is necessary . . . Nonconformist chapels are generally worse, on account of the frequency of galleries and the consequent crowding. Worst of all are probably the numerous mission rooms which, through the energy of the clergy, are found in such large numbers in the poorer districts of our large towns. These are frequently improvised out of a couple of cottages. No architect is consulted on the subject, the alterations are made by a local builder, and sanitary conditions are absolutely unthought of. The strictest economy is observed, especially in the heating apparatus, which is generally a small stove, and every Sunday a large class of more or less unwashed children is succeeded by a crowd of totally unwashed adults, till the atmosphere can only be described as sickening.[7]

[7] loc. cit.

These last observations on the 'great unwashed' are not snobbery; Jacob clearly spoke sober truth based upon personal observation. Medical practitioners, in the course of their normal rounds and as visitors accompanying inspectors of mines and factories, had unrivalled opportunities for observing the varieties of environmental disaster the nineteenth century had bred, and would be exposed to conditions that rarely came to the notice of architects. The common contempt of medical men for the inhibitions of convention, and their rationalist belief in direct physical action are well enough known to leave no reason to be surprised that they frequently went beyond mere verbal protests at the conditions of the time. Not only did they often exert political leverage at the local and national level, but some also put up exemplary structures.

In the absence of any convenient source of directly applicable environmental power, they had to apply their medical knowledge and elementary principles of environmental physics in precisely the same holistic manner as is implicit in Jacob's critiques, designing the whole structure and use of the house anew, in order to achieve their environmental aims. In Liverpool, for instance, a Dr Drysdale, and a Dr Hayward, both built houses in the 1860's whose whole design turned around problems of ventilation and heating. J. J. Drysdale was the pioneer; his Sandbourne House of 1860 still stands. John Hayward's house, the Octagon in Grove Street, is a more complex and slightly more sophisticated affair, and was completed seven years later, and also survives, though in a very dilapidated condition. Both houses are well documented, as are their designer's intentions, because the two doctors collaborated on a text-book, *Health and Comfort in House-building* (1872) while Hayward read a paper on his Grove Street house to the Liverpool Architectural and Archaeological Society shortly after its completion.

His descriptions of the form and functioning of the Octagon in these two documents are so lucid and systematic, and give so good a picture of the environmental technology available and exploited in practice, that there is little left for later scholarship to add. All that need be said, in truth, is to draw attention to the way in which the whole plan, section and construction of the house, has been affected by his determination to control the ventilation, and the matching manner in which practically everything within the house, including the gas-lighting, is consciously set to work to assist the structure in realising that aim. A compact description of the house is given in *Health and Comfort*:

> The basement is devoted principally to the collection and warming of the fresh air. On the ground floor are the cellars (*scil*. cold stores) a ball-room, two professional rooms . . . The first floor is the living floor . . . The second consists of the family bed-rooms with breakfast room . . . and the third floor of the servants' bed-rooms with children's play-room, store room and two water-cistern rooms. And above, beneath the ridge of the roof, is the foul air chamber, into which all the vitiated air of all the rooms in the house is collected, and from which it is drawn by the kitchen fire, by means of a shaft passing down to the ground floor, and then ascending behind the kitchen fire, and up the kitchen chimney stack round the smoke flue.[8]

The use of an ascending/descending convection-duct of this sort,

[8] op. cit., p 68. All the available information on the Octagon, including survey drawings (which do not detail the duct-work, however) was brought together in a joint thesis report by J. I. Chambers, A. B. Shaw, R. J. Winter and R. N. Dent, which is now in the library of the School of Architecture, Liverpool University.

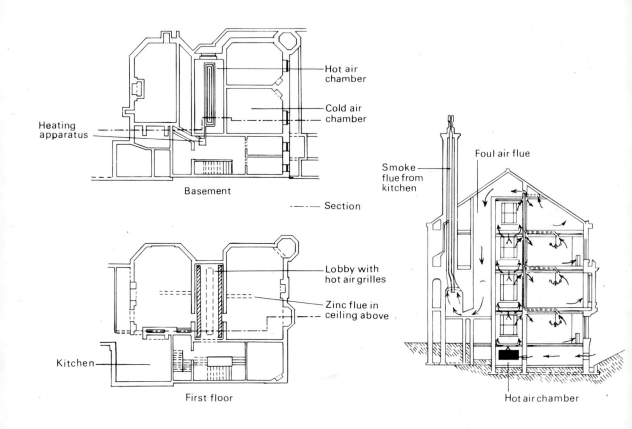

Hot air chamber

Cold air chamber

Heating apparatus

Basement

—·—— Section

Lobby with hot air grilles

Zinc flue in ceiling above

Kitchen

First floor

Smoke flue from kitchen

Foul air flue

Hot air chamber

powered by waste heat (in this case, from the ever-burning kitchen range) was a pretty common form of air extract in the era before suitable fans were available. What is uncommon in the design of the Octagon is the way in which all the principal rooms open off closed lobbies, separated by doors from the hall and staircase. These lobbies, superimposed exactly in plan, form a vertical supply duct (called a 'corridor' by Hayward) delivering cleaned and warmed

Above: plans and section of the Octagon, Grove Street, Liverpool, built for his own occupation by Dr John Hayward, 1867.

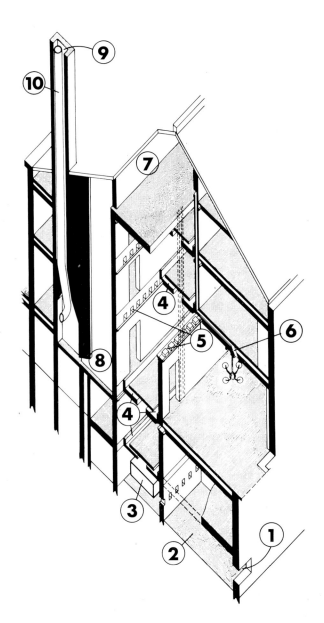

Diagrammatic cut-away perspective of the Octagon, to show the circulation of the air.

1. Fresh air intake
2. Settling chamber in basement
3. Heating coils
4. Air passages in lobby floors
5. Air passages in cornice
6. Extract above gas lamp
7. Foul air chamber
8. Foul air down duct
9. Foul air chimney
10. Flue from kitchen range

air to the rooms, in a manner described as follows, to the Architectural and Archaeological Society:

> Along the centre of the ceiling of each storey of the central corridor is an ornamental lattice-work two feet wide, and along each side of the floor above is an iron grating one foot wide, these allow the warmed air to ascend from the lobby beneath to the lobby above, but check it for the supply of each floor, and prevent it rising directly to the top one.
>
> Along beneath the ceiling of the basement of this corridor run five coils of Perkins' one-inch diameter hot water pipes. Fresh air enters into the lower part of this basement and, rising, is heated by the heated pipes, and passes through into the lobby of the ground floor, and thence into the lobby of the first floor, and thence into the lobby of the second floor and thence into that of the third floor, so that the central corridor is filled from the ground floor to the attics with fresh warmed air.
>
> . . . Out of this central corridor all the principal apartments of the house open; and out of it, and out of it only, they receive their supply of fresh air.
>
> The cornice round the ceiling of the corridor, and of each of the rooms opening out of 't, has a lattice enrichment seven inches deep, and the wall between these two cornices is perforated by as many seven-inch by five-inch openings as the joists will allow . . .
>
> Over the gasolier in the centre of each room is a perforated ornamental, covering a nine-inch square opening into a zinc tube nine inches by four-and-a-half inches . . . this zinc tube goes along between the joists of the ceiling into a nine-inch by four-and-a-half inch flue in the brickwork of the wall, between the corridor and the room above, where it is regulated by a valve. The flue rises up inside the wall and opens into the foul air chamber formed under the roof of the attic. The flue from each room *opens separately* into this chamber, and there is also a flue from the cloak room, the dressing room, the bath room, and the kitchen, and from all the water-closets, even the servants' in the basement; there are eighteen flues . . . Out of the north end of this chamber goes a brick flue or shaft, six feet by fourteen inches, taken from the back staircase . . . this outlet or shaft goes straight down to below the first floor, and then crosses eastward and rises up behind the kitchen fireplace, it is then collected in a square shaft . . . of at least *five square feet* surrounding the kitchen smoke flue, and these together form a large chimney stack, which is carried up to a greater height than any other chimney in the house, so as to secure a long siphon and a strong draught.[9]

[9] ibid., pp. 92–94.

It might appear from the above that the ventilation would be restricted, in practice, to a shallow zone below the ceiling, the air entering through the perforated cornice and leaving through the ornamental rose over the gas lamps. But this supply of warm air was not intended to heat the house—heat was provided in every room by a conventional fireplace, whose chimney would pull the fresh air down from the cornice and across the room. The function of the elaborate supply-system for fresh warmed air was the same as the reasoning behind the permanently sealed windows: to prevent cold draughts, which were the normal concomitant of any supply of fresh air in conventional structures of the period. The elaborateness of the provision that had to be made to achieve this, and the consequential effect upon the whole form and structure of the house will probably seem nowadays to be totally disproportionate to the benefit gained, but one must remember that 'the draught' was (and is) an obsessive enemy of thermal comfort in England. To have gained a draughtless air-supply, free from the common dusts and grits of the urban atmosphere (they had been allowed to settle out in the chambers of the basement) would have appeared a major gain in domestic environmental management at the time, especially when it is seen against a background of the going state of environmental knowledge and practice.

By the 1860's, the practice of heating had begun to stand less upon rule of thumb than upon a quantifiable body of knowledge of performance and control—at least the performance of boilers and radiators seems to have been within the scope of numerical expression and calculation. The progress of quantification, plus the solidification of custom, were to have a stultifying effect on speculative and humane thinking about the supply of heat, however, and by the end of the century the use of ordinary human imagination, if we are to judge by works like Baldwin's *Outline of Heating, Ventilating and Warming* (1899), had almost come to a stop. For Baldwin, it appears that questions of human comfort and physiological response to thermal stimuli either do not exist, or are not open to discussion. The aims and priorities of heating are set out in crisp phrases of crushingly mechanistic insensitivity, thus:

BOILER: in warming by steam or water, the boiler is generally the first consideration.[10]

It is usual to maintain a temperature of 70°F within a room.[11]

It may be asked 'Why is the question of condensation the first consideration?' and in reply I will say that it furnishes us with the first item of data on which to base all our other calculations . . .[12]

It may be objected that Baldwin can treat matters thus because there was a going consensus of opinion that rooms should be kept at 70°F and that no further human study was needed at the time. In fact, there was no such consensus, and never has been, though later 'environmentalists' have made equally procrustean propositions—eg., Le Corbusier's proposition to maintain a temperature of 18°C in buildings in all parts of the world irrespective of local need or preference. In any case, what makes Baldwin's approach to heating appear unattractive and mechanistic is that his observations on ventilating, which was not yet a quantifiable topic, are humane and direct. They are truly observations, based upon manifest personal experience, and they support a genuine discussion of the problems involved, without the short-hand dogmatism of his views on heating. He notes that:

The early investigators depended largely on the sense of smell as a guide to the vitiation of rooms.[13]

and he, like his immediate predecessors and most of his distinguished contemporaries, was clearly a 'nose-man' and an experimentalist in the field.

Indeed, one of the great rewards of studying the environmental literature of the late nineteenth century is that while heating had been reduced to rule and formulae, ventilating had not been, but was still open to discussion, much of it sensitive, more of it speculative. As industrialised societies fought their way out of the soot, smog and grosser pollutants of their atmosphere, they came up against a situation which clearly baffled men of common sense accustomed to the practical mechanical solution. Whereas 'heat' or 'cold' could be satisfactorily measured with relatively

[10] *Outline of Heating, Ventilating and Warming*, New York, 1899, p 22.
[11] ibid., p 34.
[12] ibid., p 32.

[13] ibid., p 13.

simple instruments and their causes identified, the 'freshness' or 'stuffiness' of air could not, largely because their causes could not be identified. Even when the two worst offenders in the vitiation of air had been finally exposed—carbon dioxide and excess humidity —neither was as susceptible to easy measurement and constant monitoring as heat, and neither was as susceptible of direct personal observation without instruments because both, in the forms normally encountered, are invisible and odourless.

Thus, the two first impacts on the human senses were normally 'the smell' and 'the draught'. The smell was observed in all inhabited interiors, especially when they were crowded and heated, the draught seemed to arise whenever those interiors were aired sufficiently to remove the smell. Attempts to do away with the draught could be as complex as Dr Hayward's, or as self-frustrating as a case recorded by Baldwin:

> In well-built modern residences the construction is often so good that it will hold water . . . a grand New York residence was so air-tight that the air to supply the grate fires had to come down the register flues (and) the air had to come down the ventilating flue of the hood of the range in order to supply the range fire, until a window was opened.[14]

[14] ibid., p 53.

However, if the draught could be stopped at source, then the smell had to be stopped at source too, and this proved difficult while the ultimate causes of vitiation and stuffiness remained odourless and therefore undetected by a generation of engineers whose ultimate arbiter of ventilation was the human nose. Precise knowledge remained fragmentary and ill-diffused, surrounded by private suppositions verging on the superstitious.

For instance, the growth of scientific knowledge and speculation about the key vitiants was, in fact, as described by Dwight Kimball in 1929, who, after setting down the primacy of the great French chemist, Lavoisier

> who in 1777 began the study of oxygen and carbon dioxide

goes on to list the succession of true pioneers after him:

> Following this for about a hundred years the carbon dioxide theory

prevailed in ventilation. Then came the theory of Max von Pettenkofer (1862–3) who first established the conclusion that bad ventilation should be charged to other factors than carbon dioxide. The harmful effects of bad air and the beneficial effects of good air later led to the erroneous theory of hypothetical organic substances in the air. Then came the recognition of the work of Hermans (1883), Flügge (1905) and Hill (1913), proving that the thermal, rather than the chemical properties of air are of vital importance in connection with ventilation, insofar as normally occupied spaces are concerned.[15]

Not only had those 'thermal properties' (as measured by the wet and dry bulb thermometer) proven extremely elusive, but the pioneers had been misunderstood. Max von Pettenkofer, because he had proposed the measure of carbon dioxide as a workable guide to the level of all pollutants, was mistaken for a proponent of the carbon dioxide theory, which he manifestly was not. If a man of Pettenkofer's eminence and fame (he was the father of modern hygiene as we know it) could be misunderstood, the general confusion of knowledge should cause no surprise. Solid and responsible practical men stood upon their private experience (having nothing else to rely upon) even to combat their mistaken image of Pettenkofer, and in the process revealed the enormity and nausea of the olfactory problem. Thus Konrad Meier, a New York heating consultant, in his *Reflections on Heating and Ventilating Engineering* (1904) wrote:

> Carbonic acid is not a poison in the ordinary sense of the word, and much larger quantities than generally assumed may be present without causing ill-effect . . . On the other hand, substances and impurities that cannot be estimated from the presence of carbonic acid, as for instance an excessive amount of vapour of water, sickly odours from respiratory organs, unclean teeth, perspiration, untidy clothing, the presence of microbes due to various conditions, stuffy air from dusty carpets and draperies, and many other factors that may combine, will in most cases cause greater discomfort and greater ill-health.[16]

What is striking about Meier's demonology of bad air is that it not only includes the real culprits, carbon dioxide (carbonic acid gas) and water vapour, but still retains nearly all the common Victorian villains, such as 'sickly odours' and makes provision for

[15] in *Heating, Piping and Air Conditioning*, June 1929, 'Air-conditioning, its future in the field of human comfort', p 93.

[16] *Reflections . . . etc.*, p 20. This document, found among the vast deposit of technical pamphlets that have come to rest in the New York Public Library (Bound Pamphlets, VEW, pv12, No. 1) appears from its format and content to have been some sort of annual address to the New York Branch of the American Society of Mechanical Engineers.

any demons he may have overlooked ('many other factors'). This was how he had observed the situation according to the evidence of his own nose. The pre-occupation with body odours may strike modern readers as a trifle obsessive and neurotic, but so general and emphatic was the apparent nasal evidence on this subject that the belief in an organic poison, mentioned by Kimball (above), is understandable—what else could have caused the unmistakable and ever-present 'stuffy smell' in occupied interiors, especially at a time when the knowledge of air-borne bacteria was beginning to diffuse among the general public?

But the point about 'the smell' in this sense is that it was not a gross industrial pollutant that caused it, but the mere presence of breathing human beings in a closed space. The better those spaces were closed by improved construction, the better lit by gas and the better heated, the worse the situation became, and it was not something that could be bettered by social legislation or moving out into the country. In other words it was not—like working in a mine or factory, or living in a tenement—a hazard that the educated and well-to-do could avoid by their usual methods. To the factors bearing upon environmental reform, the considerations of hygiene and efficiency, economy and profit, already cited, must be added the aesthetic distaste of wellbred persons for the stuffiness of their interiors and the consequent head-aches with which they woke so often in the morning (Professor Jacob also notes that clergymen had Monday head-aches after a full Sunday stint in crowded churches).

These people, in households that bred, or were presided over by, 'New Women' or their emancipated equivalents in non-Anglo-saxon countries, were the main support and proving ground for any environmental innovations that could be produced in domestic sized packages. The rise of electric lighting is inseparable from this milieu, its cleanliness and slightly mysterious quality seemed to chime in well with the interests of an intelligentsia that was turning away from the gross materialism and determinism that had characterised so many mid-century attitudes, in favour of a more mystical and aesthetic approach. The tenuous curves, pale walls and luminous decorations of Art Nouveau and Tiffany would be unthinkable without electric lighting, not only in the purely physical sense that effluents of gas lighting would have rotted delicate fabrics and darkened the decor, but also in the purely aesthetic sense that the quality and distribution of light that could be achieved is entirely apt to the style.

Insofar as Art Nouveau is the first of the new styles and not the last of the old, it is in its determination to repudiate the norms of nineteenth-century interior design, including its environmental standards. Art and technology combined to reject the dark, the coarse, the overstuffed and the stuffy. There had been previous attempts, before the 1890's, but they had been relatively inconsequential in the absence of a fundamental revolution in environmental technique. But if the sudden availability of electric lighting marks the turning point in that revolution, the ferment of improvement and innovation had been going on for most of the century. The evolution of the kit of parts needed to revolutionise the environment of men is a history in itself.

4. The kit of parts: heat and light

The preceding chapter will already have given some idea of the kind of technology of environment that was becoming available during the nineteenth century. The development of the art needs to be discussed in somewhat fuller detail, however, even though there is no intention of providing a complete technological history within the compass of the present work. Chiefly, it is important to establish the changes in the type of environmental power that could be delivered into an inhabited space. In the middle of the nineteenth century, the nature of that power was still essentially primitive, its basic characteristic was that fuel was burned more or less at the point where power had to be applied—coal or wood in grates and boilers, oil, gas or tallow in lamps and candles. In the absence of machinery of domestic scale most of this power had, of need, to be applied directly and crudely to the immediate environment, since water was the only substance commonly channelled through pipes or conduits.

However, the fact that water could be heated, and then circulated through pipes, afforded the prototype of most later forms of sophisticated environmental control—the combustion of the fuel at one convenient point, and the application of the energy thus generated at some other convenient or necessary point. The first proposals to use hot water in this way go back into the Renaissance, but their practical application belongs to the pioneer phases of steam technology in the late eighteenth century—James Watt had his own office heated by steam in 1784, and legend has it that the earliest building heated from the first by such methods was Matthew Murray's 'Steam Hall' in Leeds in the first years of the nineteenth century.

Given boilers of moderate efficiency, economically produced

heat could be circulated by convection, without use of pumps, through fairly complex networks of piping to suitably placed radiators, and its comparative simplicity made it a practicable proposition for domestic installation. With the addition of pumped circulation and other refinements, the basic technology could be adapted to much larger installations, provided there was sufficient janitorial skill to operate them. By the 1860's, heating by steam or hot water could be looked for in most buildings, public or domestic, of any pretensions; considerable skill had accumulated in the design of the installations, both on paper and at the level of field decisions that had to be made by foremen and pipe-fitters. This was one of the great reservoirs of skills on which much of the environmental revolution was founded, though there is some evidence that the persistence of drawing office habits and fitters' folkways may have ballasted down the aspirations of innovators on occasions—Willis Carrier on one occasion had to correct the operating habits of an engineman before the cooling plant of one of his early air-conditioning installations would operate properly.

But piped steam heating is also, with the electric telegraph, the prototype of another—obvious and necessary—development in the use of environmental power. If heat could be distributed from a central boiler to different parts of the house, it could also be distributed to different houses

It is doubtless true that in the early days of steam heating, various people have heated more than one building from a single source. However, just as Thomas A. Edison is looked upon as the father of the central lighting station, so in the heating industry there is one man generally named as the pioneer of central station heating, Mr Birdsill Holly, of Lockport, NY.

In 1876, Mr Holly ran an underground line from a boiler in his residence to a barn at the rear of his property and later connected an adjoining house. In 1877 he constructed his first experimental plant at Lockport, in the state of New York, and a number of residences, stores and offices were successfully heated during the following winter.[1]

Although Holly was not really an innovator of the same order as Edison, there is some limited justice in the comparison. Both finally

[1] Bushnell and Orr, *District Heating*, New York, 1915, p 2.

went 'on stream' in downtown New York City in 1882–1883, supplying a basically similar service: clean environmental power from a central source. Whereas previous technologies had supplied the householder with raw or partly processed fuel (eg., coal gas) to be more or less inconveniently or messily burned in the room, Holly was supplying clean and directly usable heat that left no residue in the house to be cleaned up, and consumed none of the air available.

The elimination, in the process, of the open flame, is a development of some consequence, which will be discussed later. The *one of source of garbage* next topic to concern us here is the application of heat, however supplied or generated, to the interior. In general, the technology of the mid-nineteenth century could offer little—at a domestic scale—beyond letting the heat find its own way into the environment by simple radiation and convection. However much the innumerable patented 'improvements' to stoves in the nineteenth century might have boosted their performance by better combustion or transference of heat to the ambient air, however much the design of steam and hot water radiators may have become sophisticated, the stove, grate or radiator stood at some point in the room dictated by custom, convenience or aesthetic preference and the warming of the space around it was at the mercy of draughts, open windows, local convection from lamps, obstruction due to furniture, etc. One can almost say that the only serious attempt to cope with this situation that achieved any widespread distribution is the inglenook, virtually a room within the room, around the fireplaces of rooms in large houses by Frank Lloyd Wright, C. F. A. Voysey and their contemporaries. These screened areas with built-in seats provided an area of reliable thermal performance, shielded from draughts. Though ultimately a revival of mediaeval usage, they could only be properly revived in a epoch that already disposed of piped central heating—the effect of trapping so much of the heat available in an area around the fire would have been to deprive the rest of the room thermally, were background heat not available.

47

However, the improvements in heat transfer to which reference has been made, were not negligible, and were based, in most cases, on the separation of the convecting warmed air from any smoke or fumes that must be disposed of. There seems little doubt who was the true father of this art.

> The principle of heating a room with warm air was introduced by Benjamin Franklin in 1742. His stove of that date contained a chamber surrounded by iron plates and fed by a cold air box, openings for the escape of the air being in the sides or jambs at the top of the chamber.

wrote William Gage Snow in 1923, and added

> The warm air furnace of today is identical in principle but more elaborated.[2]

Other developments of the Franklin's—or the immediately succeeding—generation, included aids to more efficient combustion, such as Rumford's restricted throat grate. Many improved stoves and grates also called for a separated supply of combustion air, drawn from outside the space to be heated. This growing sophistication in the handling of air, carried further by such techniques as drawing it in from the outside only through grilles containing, or serving, radiators, was rendered necessary by the steady reduction of sources of accidental ventilation, due to the better sealing of windows, for instance, or the disappearance of the chimney in spaces where direct combustion was not the source of heat. As less was left to accident, more aspects of the thermal and ventilation performance of buildings had to be consciously controlled and investigated—Jacob, for instance, is able to show diagrams of air movement and heat distribution based upon controlled tests.

But, in all these improvements and innovations, the most consequential is perhaps the separation of the combustion gases (both input and output) from the air warming the room.[3] Once the heating air was on a different circuit to that serving and produced by combustion, that independence could be exploited. The circuit could include rooms other than that in which the stove was located, if suitable openings in walls, or ducts, were provided. Sooner or

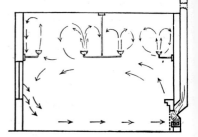

Air-circulation in a room heated by an open fire, and lit by gas.

[2] *Furnace Heating*, New York, 6th ed., 1923, p 213.

[3] Hot air heating is sometimes spoken of as senior to heating by hot water or steam, usually on the evidence of Roman hypocausts, and other primitive systems. Since such under-floor arrangements circulated the products of combustion promiscuously with the hot air, they clearly do not fall into the class of modern hot air systems, descended from the Franklin stove, discussed in this chapter, and their involvement with modern architecture since 1850 seems negligible.

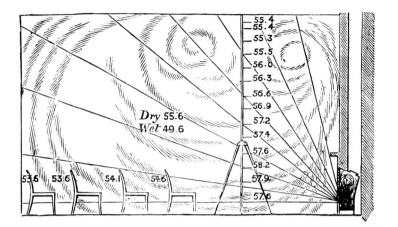

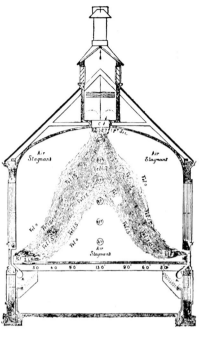

later someone had to dismiss the stove to the basement, tap the air output from its hot air box, and duct this heated air to the parts of the house where heat was needed.

Although this was to prove a portentous invention, since it established the basic heating method for most North American residences, its origins seem already to be lost even beyond the reach of legend. William Gage Snow quoted[4] the following from an unspecified issue of *The Metal Worker*:

> . . . just who was the first to improvise this heating apparatus or when it was done, is difficult to learn. The date, while it cannot be fixed with certainty, is in all probability prior to 1836. There is an impression among many of the older hot air furnacemen that experiments in this line were numerous in the vicinity of Hartford, Conn., and that along about 1840 a number of hot air stoves are known to have come into existence.

It should be noted that the use of the hot air stove or furnace brought with it an added benefit over and above heating—since it delivered the heat by means of air, and only when that air moved, it was inseparable from ventilation. Either the movement of the hot air improved the ventilation, or the ventilation had to be improved to the point where the hot air could move. But this was not always easy to achieve in conventional architectural formats.

Left: air movement, temperature distribution and humidity in a lecture-room, as measured by Campbell in 1857.
Above: currents of air in a 'model' hospital ward.
(both these diagrams reprinted by Professor Jacob from Galton's *Healthy Homes*, 1880.)

[4] loc. cit.

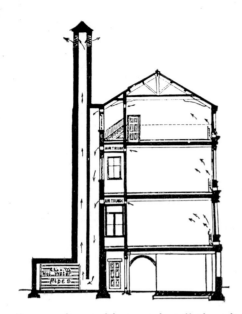

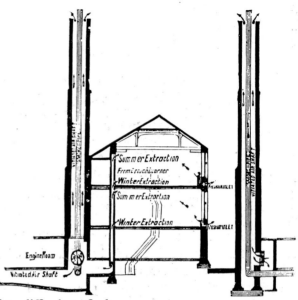

In complex multi-storey installations it was often difficult to find space for vertical riser ducts in conventional Victorian construction—though ingenious use of the furred spaces behind apsidal and polygonal room ends seems to have been made in the Boston area at least. But where ducted hot air finally came into its own was in the nearly-standardised single storey houses with basements that have spread from the middle-west to almost every part of North America. The basement provided not only room for the furnace but also freedom to distribute the ductwork efficiently and economically, and thus to deliver the warmed air to the places where it is needed most, around the perimeter of the house.

In the mid-nineteenth century, however, when the problems of warming and ventilation were being tackled together for the first time in terms of conscious design, the price of efficiency was usually the adaptation of the whole structure to the needs of convected air circulation—on a small scale, in the manner of the Octagon in Liverpool, discussed in the previous chapter, on a large scale in the

Heating and ventilating with thermal siphon extract, left, and with powered fan extract, right. Below: mixing valve with remote control for domestic hot air distribution, Sturtevant catalogue, 1906.

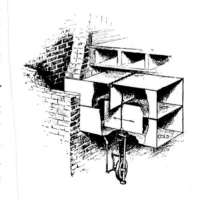

Sturtevant installation to provide warm air at the entrances and display-areas of a store in Boston, from the 1906 catalogue.

manner of giant brick ducts, often with their own source of heat to stimulate air-movement, which dragged air through institutional and civic buildings. Though considerable achievements were wrought with these techniques, at the expense of inconveniences in plan and section, the arts of both ventilation and heating really waited upon the development of effective blowing fans.

William Gage Snow records the 'embryo idea of a fan furnace' in a B. F. Sturtevant Company catalogue in 1860 (the year of that celebrated ventilating-company's foundation). But the idea goes back much further, of course. J. T. Desagulier invented the very term *ventilator* to describe the man who turned the crank of the centrifugal fans he was proposing, to supply air to the lower decks of naval vessels and the chamber of the English House of Commons in 1736. Nevertheless it was in the period after 1860 that fan-forced ventilation began to flourish. The pressing needs of mining and shipping, of industrial processing (such as the drying of tea, for which Davidson developed his *Sirocco* fans) and the increasing size and complexity of buildings all provided powerful stimuli to invention; the steam engine and, later, the slow running gas-engine drawing on the common town gas mains, provided the power. By 1870, the Sturtevant Company could patent a steam-coils-plus-centrifugal fan combination that was well out of any 'embryo' stage.[5]

[5] chronology in M. Ingels, *Willis Carrier, Father of Air-conditioning*, Garden City, 1952.

But the size and weight of such plant often made its location within the building-structure difficult, so that conservative ventilating experts could argue as late as 1882

> To attempt to draw down the foul air from the upper storeys of a building and to conduct it by an underground channel to an engine shaft, is generally a very roundabout and unscientific mode of ventilation . . . On the whole, it has been pronounced by competent men, that the heated shaft has more in its favour than the fan driven by a steam-engine . . .[6]

[6] *The Building News*, June 9, 1882, 'A Note on Hospital Ventilation', p 709.

The problem was to be resolved by finding other places than the conventional basement for the location of the fans and their attendant plant, or by making the plant less bulky and massive, or by using the fans in a different way, as in the Plenum system and its derivatives, in which force fans were used to keep the ventilated volumes under a slight pressure, so that foul air would find its own way out through accidental or designed exits (sanitary areas were a preferred outlet-route for systems of this sort, since the air-flow would carry the dreaded smell of 'drains' out of the building directly).

But the progressive application of fans was held back, until the last years of the century, by two major factors. One of these obstructions—lack of aerodynamic knowledge—was worn away slowly, by the accumulated practical experience of companies like Sturtevant and Sirocco; or by the design-work of men like Rateau in France and the theories of Joukovsky in Russia—the one

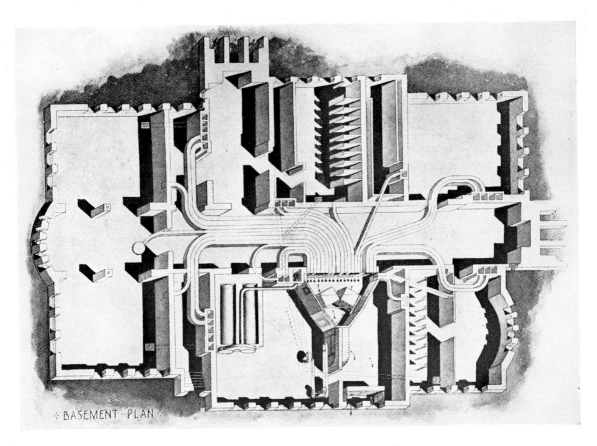

Ducting, boiler, fan and heating chamber in the basement of a school in Menominee, Mich., Sturtevant catalogue, 1906.

·BASEMENT·PLAN·

the inventor of the modern high-speed centrifugal fan, the other the intellectual parent of axial flow fans. The other obstruction to progress was the lack of a small power source adaptable to fans of domestic or personal scale, and here the breakthrough seems to have been more sudden, waiting upon the almost simultaneous development of domestic electrification and of Nikola Tesla's alternating-current motors. Both these basic developments belong to the 1880's, so do the first mentions of electric fans as room-coolers in downtown New York. The relative smallness of plant

brought the simple fan

53

achieved and its presence in the actual room being ventilated, is probably less important than the range of descending sizes of electric power-unit and ventilator and the great handiness of their management and location. It was upon these bases that the growing sophistication of ventilating techniques in the twentieth century was to be built.

Until such time, however, ventilation-technology had to make do with, fundamentally, the same kit of parts as Dr Hayward had used, and since this normally involved the direct application of heat as a source of convecting power, its strong point was emphatically not summer cooling. Although something could be done by stacking ice in intake ducts, or (towards the end of the century) by the use of cooled coils supplied by a refrigerating plant to chill intake air, the mere dropping of the temperature did not necessarily promote comfort, since the process might raise the relative humidity. As late as 1906, the ingenious heating/cooling plant devised by A. M. Feldman for the Kuhn and Loeb bank in New York (its ingenuities were as much architectural as mechanical, and will be discussed in a later chapter) while it could pull the temperature in the banking hall down to a figure ten degrees lower than an external shade temperature of 91°F, did so at the cost of boosting the relative humidity from 53% to 63%.

For most of those locations (hot humid areas) where cooling was felt to be necessary, humidity control was equally necessary— what was needed was the sort of total environmental control that only full air-conditioning could supply. But there was probably little point in even attempting total control until the atmosphere had been cleansed *at source* of its worst and most persistent class of indoor pollutants, the waste products of combustion from illumination-fuels. Interactions between the controls of the luminous and atmospheric environments are almost inevitable if only because of the heat-load imposed by illumination-sources, as in the 1940's, when better air-conditioning and the reduced thermal output of fluorescent lamps gave new freedom in the design of office-blocks.

But, given the much greater heat load of flame light sources, and the atmospheric load of water-vapour, carbon oxides and pure carbon which they also generated, there would seem to have been little point in Carrier and Cramer even starting on air-conditioning until the filament electric lamp had abolished most of this atmospheric garbage at a single blow. The rise of air-conditioning can conveniently wait till chapter 9, but the revolution in illumination cannot, so fundamental is it to the attainment of the kind of environmental conditions thought proper to modern architecture.

The utilisation of artificial light rose sharply after the middle of the nineteenth century, and the increase is even more striking if measured in candle-power hours than in expenditure on fuel burned. Up till the mid-century, it is doubtful whether the illumination of the average household rose much above an almost mediaeval level: a single candle burned for an hour or two each evening, the life of the household tailored to make best possible use of the exiguous light available—that is, those with the greatest need, as in reading or sewing, closest to the lamp on the table, those with less need further away, almost a camp-fire situation, in which space was, for the moment, focused around the lamp as much as framed by the walls of the room. More efficient oil-burning lamps, such as the Argand, did not materially affect this situation, and it was the rising availability of coal gas from the mains after the middle of the century that really began the increase of fuel burned, light used, and number of lighting outlets employed. Where figures have been analysed, they can show as much as a twenty-fold increase in the actual amount of illumination employed in an average household in a city like Philadelphia, between 1855 and 1895.[7]

The sheer amount of light available and used, in itself must constitute a major revolution in human life; the means of obtaining that light remained prehistoric, piped gas notwithstanding. All the increase noted above, barring the last five years or so to 1895, was obtained by means of open flames, inefficiently operated.

[7] figures given by Dr Walton Clark in *NEL.* *lletin*, 1910, Vol. X (III, new series, No. 10).

Inefficient they had to be, since the actual source of light was the incandescence of unburned carbon particles in the flame, whether that flame was fuelled by oil or gas, whether it burned in a fish-tail, bat's-wing or any other type of burner. Having served as the medium of incandescence, the carbon particles then ascended in a narrow column of soot and deposited themselves on the ceiling, as much as anywhere else. It was to deal with this encrustation of soot, which left the ceiling dark grey, or even black, above the gasolier and shading off to lighter greys at the cornice, that nineteenth-century house-keepers elaborated the ritual of spring cleaning. At the end of the soot-generating lighting season (autumn–winter) they took all soot-gathering fabrics, draperies, carpets and up-holstery out of the room and at least beat the loose soot out of them, and at the same time the ceiling could be cleaned or even re-whitened. But such domestic upheavals grew less and less welcome even in households that were still accustomed to total disorganisa-tion every Monday in order to accommodate the equally elaborate ritual of 'washday'. A clean light source could clearly do much to reduce the rigours of both rituals, and even the increasing amount of dirty light in use was increasingly revealing a matching increase in general domestic dirt and pollution.

Attempts to bring the existing illumination-products within bounds varied, but most were concentrated—understandably—on the area above the gasolier itself. The use of extract grilles above the light-fitting was not peculiar to the Octagon, Liverpool; in practice, it was a method both of disposing of sooty wastes and of exploiting the thermal waste to convect foul air out from the heavily polluted zone immediately under the ceiling. Such grilles were usually incorporated in an ornamental or rose—a sizable plate of foliate or architectural decoration in fairly heavy relief, surrounding the point of suspension of the light-fitting. The ripe decoration which characterised such objects may or may not have been there primarily at the behest of ripe Victorian tastes in decorative art, but the depth of its relief, its undercuttings and convolutions, also appear to have done much to trap sooty waste within the confines of the ornamentation, and discourage it from spreading right across the ceiling.

However, the worst of the problem was suddenly avoided by a major breakthrough in gas-lighting technology at the beginning of the 1880's, when the egregious Austrian inventor, Baron Auer von Welsbach produced a commercially viable gas mantle—that is, a bulb of fireproof fabric impregnated with oxides of rare earths which would incandesce in the heat of a gas flame. Since the mantle itself incandesced, there was no need for unburned carbon to do so, and the flame could burn efficiently on the principle of the Bunsen burner, with its correctly regulated flow of air. As a result, the output of sooty wastes was greatly reduced (though rarely eliminated under normal domestic conditions) even though the output of heat remained considerable.

This was an enormous step forward, especially when the neat and convenient inverted mantle was introduced, and everyone who has lived with domestic lighting by gas will know that it has much to recommend it—a warm, murmuring, friendly radiance, of quite a pleasant colour-spectrum when correctly trimmed. The gas mantle, together with its heating partner, the incandescent gas fire (models using asbestos string as the radiants were available from the early 1880's) might have had a great future, but for two things. The first was von Welsbach's attitude as primary patent holder, combining as it did an almost feudal conception of absolute property rights, a Levantine deviousness in financial methods, and a straightforward nineteenth-century determination to make as big a killing as possible, which all combined to leave him trying to hold the market to ransom at the very moment when deliverance was at hand in the shape of the second thing—the perfection of a workable system of domestic electric lighting. The Welsbach mantle appeared on the scene just too late to establish itself fully before the whole basis of gas illumination was swept away by the triumph of Edison and Swan.

Electric lighting offered in a single package, a double solution to the environmental problems posed by gas; it generated less heat, and made no soot. Further it needed dramatically less servicing and trimming than gas, and could be installed in many restricted spaces where gas with its heat and need for air would have been barely practicable. Given these advantages, electric lighting was irresistible, however much more expensive than gas in installation-cost and running consumption it might have been at first. So attractive was it, that hard headed and cost-conscious business men called for its installation in new buildings even before the supply of electric power was available. A justifiably famous case was that of the Montauk block in Chicago, designed by Burnham and Root. In a letter to their Chicago agent, Owen F. Alldis, dated February 5, 1881 (a clear twelve-month before public mains supply of electricity was available anywhere in the world) the proprietors suggested:

> The less plumbing the less trouble. It should be concentrated as much as possible, all pipes to show and be accessible, including gas-pipes. It might also be advisable to put in wires for future electric lights. It is not uncommon to do it in Boston now.[8]

If even such tough commercial minds could be captivated in this way by a service which was still no more than a promise, we can hardly be surprised that the advent of electricity as a source of lighting and environmental power was awaited with something like religious awe, as if men had been vouchsafed a vision of beneficent magic. In May of 1882, the *annus mirabilis* of the incandescent electric lamp, John Slater, a Fellow of the Royal Institute of British Architects, read a paper to the Institute on 'Recent Progress in the Electric Lighting of Buildings.' It was a major occasion; the room

> . . . was lighted by incandescent lamps of the Swan, Edison, Lane-Fox and Maxim types, supplied with power from an accumulator invented by Messrs Sellon and Voelkmar, and which stood in the room . . .[9]

[8] quoted in C. Condit, *The Chicago School of Architecture*, Chicago, 1964, p 53.

[9] reported in *The Building News*, May 19, pp 600*off*. The name of John Slater does not figure very large in the annals of British

and thus surrounded and illuminated by visible proofs of this seemingly miraculous light source, Slater observed that the

> . . . revolution was mainly due to the invention of incandescent lighting . . . a stable and unchanging point of light, in contradistinction to the arc-light, in which the glowing material was continually disintegrating and burning away.
>
> . . . the economic value of the new means of illumination had if anything, been underrated. The readiness with which the incandescent bulbs lent themselves to any scheme of decoration was one of their chief attractions. It would be undesirable to follow the lines of gas fittings, as the conditions were so completely altered, but points of light could be placed wherever they were required, and there was no fear of blackening ceilings, or of setting fire to the most easily ignited materials. The progress of this system of lighting had been so rapid that architects had as yet had no time to turn their attention to its decorative capabilities, but when they did so they would find it fulfil every requirement for perfect lighting.[10]

Slater's use of the word *decorative* in this passage need not be taken to mean anything merely superficial; it is clear from the rest of his text that he sensed a profound revolution in the nature and use of the built environment, even if he did not quite dispose of the vocabulary for discussing such matters that is available today. No doubt, it is this sense of a profound revolution that accounts for the almost religious solemnity of his concluding paragraph:

> The progress of electrical science is the most striking feature of the latter part of this nineteenth century, and the day is not far distant when we shall find a certain acquaintance with the subject of electrical science a necessity for us architects in our everyday work, unless we wish to be entirely in the hands of the men we employ. Science has captured the lightning it is true, but it is scarcely tamed yet; let us beware that we do not attempt to deal with this new servant ignorantly. Electricity is a new power given into our hands to work out, and it behoves us to study its nature and advantages and to guard against its risks and dangers, and learn to use it, with the older means at our disposal, in accordance with the maxim 'Usui civium, decori urbium.'[11]

The only point where one may find fault with Slater, given historical hindsight, is in his insistence on the crucial nature of the invention of the incandescent bulb. One must admit that direct

architecture, though the partnerships of his son (Slater and Moberley; Slater, Uren and Pike) played a respectable part in the twentieth century. Slater's electrical interests were not without their rewards; he rebuilt the first, temporary generating station of the Kensington network in 1895, and built other stations at Notting Hill and Wood Lane. He also built a house for Colonel Crompton, promoter of these schemes. Died, 1924.

[10] *Recent Progress* . . . etc., loc. cit.

[11] ibid.

personal acquaintance with the light sources then available would show so great a contrast between the flare of gas and the steady cool light of electricity, that the latter would be sure of having great impact—the author recalls with what vividness one of his school science teachers, then nearing retirement age, remembered as a boy seeing his first electric lamp, burning in the bottom of a tank of gold-fish in a shop window in Dublin! Nevertheless, hindsight and a suitable vocabulary now enable us to identify what we should call a triumph of systems engineering as the crucial invention. The lamp-bulb itself had been on the point of successful operation for some time—in England, Swan had a primitive paper-filament type of bulb as a laboratory toy as early as 1848, though difficulty in obtaining and securing a sufficiently good vacuum inside the bulb caused the filament to burn away too quickly for the lamp to be of any practical use. Then, towards 1877, the use of the Sprengle vacuum pump made the attainment of much harder vacuums possible, and the use of platinum (which has a coefficient of expansion very close to that of glass) for the leads through the bulb made those hard vacuums easier to seal in permanently. The carbon filaments of Swan and Edison were so nearly exactly contemporary that patent litigation between them ended in a drawn match, and they formed a joint company to exploit the British market. From 1878 onwards, a succession of patents relating to lamp bulbs and to metallic filaments in particular, appear almost annually.

But the unique achievement that makes Thomas Alva Edison the true father of electric lighting has less to do with the practicable lamp bulb (though he and his commercial backers would have been totally frustrated without one) than with his invention and assembly of the complete system to supply that lamp with commercially profitable electricity. The story of what happened between the formation of the Edison Company in 1878 and the opening of the first public mains supply in 1882 is too involved to be pursued here in detail. The outstanding point to be noticed, however, is that very few of the component parts of the system were totally original inventions. It was the conception of their total functioning together as a practicable kit for generating, controlling, measuring, distributing and utilizing power derived from a central generating station that is the great invention.

To achieve it, many detailed triumphs and felicities of technological ingenuity were required, sometimes combined with gratifying civic foresight, as when Edison, observing

> . . . you don't lift waterpipes and gas-pipes up on stilts.[12]

insisted on saddling himself with the task of inventing satisfactory and properly insulated underground conductors for his power, instead of imitating the overhead cables of telephone practice, which used the surrounding air as a cheap insulator.

Nevertheless, telephone practice, in various ways, played a large part in Edison's triumph. For a start, much of the necessary electromechanical skill needed to make and operate his system could only come from the reservoir of trained talent that had accumulated in the telegraph system of the US since the 1840's, and in the telephone system since the late 1870's. But it was out of this pool of talent (including its most talented member, Edison himself) that there came the solution of the problem that was supposed to make the supply of varying amounts of electricity to independently-minded domestic consumers as good as impossible.

As late as 1879, evidence had been given before a select committee of the House of Commons

> by such scientists as Sir William Thompson and Professor Tyndall as to the impracticability of subdividing the electric light for domestic use . . .[13]

but Edison brought with him from his years as a telegraph operator, professional craft in the routing and re-routing of small electric currents through complex networks, by means of rule of thumb techniques such as 'borrowing current' or downright underhand ones such as stealing it. He seems to have known, on the basis of

[12] cited in *Thirty Years of New York*, a promotional history published by the Edison Co., New York, 1913.

[13] Slater, loc. cit.

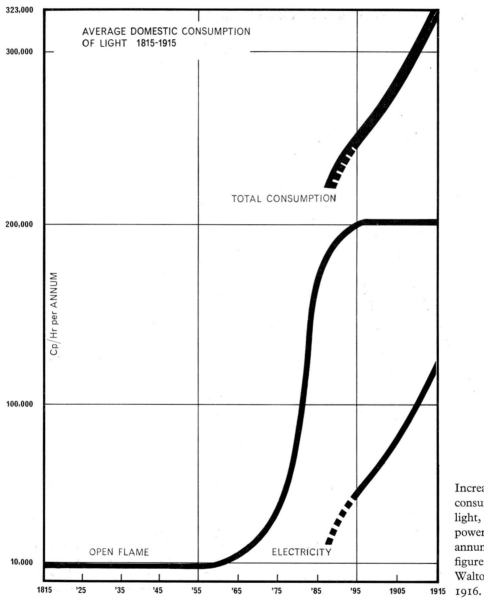

AVERAGE DOMESTIC CONSUMPTION
OF LIGHT 1815-1915

TOTAL CONSUMPTION

Cp/Hr per ANNUM

OPEN FLAME

ELECTRICITY

323,000
300,000
200,000
100,000
10,000

1815 '25 '35 '45 '55 '65 '75 '85 '95 1905 1915

Increasing domestic
consumption of
light, in candle-
power-hours per
annum, based on
figures given by Dr
Walton Clark in
1916.

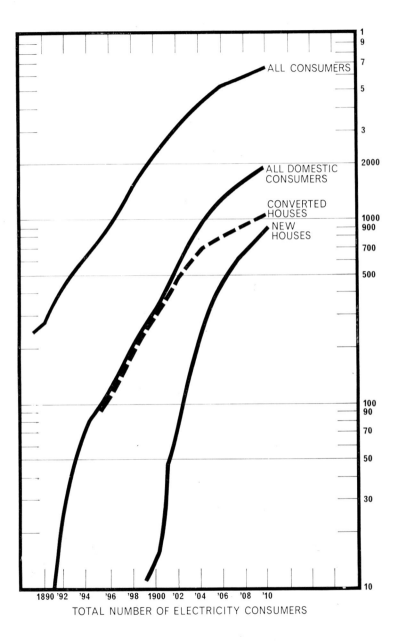

ALL CONSUMERS

ALL DOMESTIC
CONSUMERS

CONVERTED
HOUSES

NEW
HOUSES

TOTAL NUMBER OF ELECTRICITY CONSUMERS

Rate of installation of domestic
electric lighting in part of
Liverpool, 1890–1910, based on
research by H. C. Morton.

63

this experience, that although an electrical distribution network could not have the storage capacity that enables a gas or water network to cope instantly with a tap turned on or off, there were still margins enough of tolerance in a complex electrical system for suitable equipment and control techniques to handle what was theoretically beyond control, and thus to achieve an effective subdivision of the electric light.

Given this, and a method for measuring the current consumed (originally by periodically weighing the plates of an electrolytic cell) a commercial supply from a central generating station could begin. In January 1882 an Edison combined street-lighting and domestic-mains system went into operation in the area of the newly-built Holborn Viaduct area of London, some months before any of his US plants were on stream. But more than months elapsed before any similar schemes appeared in other parts of London, for the gas industry's parliamentary lobby effectively blocked any enabling legislation for the supply of electricity through cables laid in trenches in the public street—Holborn Viaduct was a legal anomaly because it was largely made-up ground —until legislation was finally pushed through in 1887, and a domestic supply was established in Kensington.

Meanwhile, the first public supply areas in the US were established in August and September of 1882, including the famous Pearl Street district in the business area of New York, thus launching the long and stormy relationship between that city and the Edison, later Consolidated-Edison, Company. Also launched by these actions was the greatest environmental revolution in human history since the domestication of fire.

The dizzy rise of the consumption of light was resumed as steeply as ever, even though the market for gas lighting went into a gradual decline. The installation of electric wiring and lamps became a branch of the construction industry that flourished even through periodical slumps that affected the rest of the business badly, right up to the First World War, because of the backlog of existing buildings ripe for conversion to electric light.[14] And over and above the new clean light source, electrification also opened the way for a host of other environmental services and domestic conveniences.

The use of domestically scaled electric fans has already been noticed, above. By 1900, manufacturers' catalogues listed and illustrated most of the cooking vessels (kettles, skillets, etc.) with built-in heating elements that are with us today, albeit in rather primitive forms, also toasters, roasters, hot-plates and ovens, radiant-panel heaters, convecting heaters, coffee-grinders, immersion heaters, and such period hardware as electric cigar-lighters and curling-tong heaters. By the time of the General Electric catalogue of 1906, the most prized of all domestic electric equipment, the electric flat-iron, was well established, as was the electric coffee-percolator and the fore-runner of the electric blanket heater. The domestic refrigerator and the vacuum cleaner were to wait until after the War for their full domestication; versions of the vacuum cleaner existed from soon after 1900, but the first Kelvinator was not sold until 1918. But captivating or necessary as all these devices might appear, none had such overwhelming advantages as electrical lighting, and none posed quite such subtle and unexpected problems for the architect and interior designer—so subtle and unexpected that they deserve a sub-chapter of their own.

*   *   *

At first, the fascination of the new clean light was such that further thought seemed unnecessary. It was enough to let the beautiful bright light flood over the interior, often from the old gas fitting, hastily modified, and increase its sheer quantity year by year as the industry obligingly supplied bigger and better bulbs—a high rate of technical improvement continued into the nineteen-teens, and made bigger and bigger outputs possible. Furthermore, the

[14] as is made clear by H. C. Morton's account of electrical connections and installations in Liverpool in the last decade of the nineteenth century. (Unpublished Master's dissertation, *A Technical Study of Liverpool Housing 1760–1938*, submitted in 1967.)

continual improvement in prism-cut glass shades (originally developed for Welsbach gas-mantle sources) made it possible to direct more and more of the available light downwards where it was wanted, without wasting a single precious candle-power. And it was to be some time before the sense or desirability of these unthinking methods were effectively questioned.

Nevertheless, from Slater onwards, there were solitary voices raised against the mere flooding of interiors with unrestrained light. Not only did he envisage the distribution of small point sources around the room, but also the use of indirect lighting from behind baffles that were part of the permanent architecture of the room (chiefly as a way of taming the tremendous brilliance of arc-lamps). But the trade and, with it, common practice, seemed committed to the central ceiling fixture, clearly in view. Some arguments can obviously be advanced for this usage. In many cases the electric light inherited the location, ceiling rose and even the piping (now used as cable conduits) of a previous gas installation (which had had to be centred in the ceiling to reduce staining of the walls). Furthermore, in rooms of unspecific function, as so many domestic rooms had to be in real life, if not in architects' visions, a central location recommends itself as a painless compromise solution to the problem of where best to put the light-source.

The interests of the trade seemed to have been two-fold; firstly, the almost annual boost in power was already beginning to produce lamps so strong and so hot that they seemed safer in the middle of the ceiling—but also, a distributive solution meant smaller lamps, and therefore no incentive to produce a more powerful model next year, so that this was one of those self-sustaining vicious spirals to which the technologies of market economies are so often prone. And again, big central fixtures were more expensive to manufacture, and thus promised a bigger mark-up to the retailer who sold them. Again, they were often so heavy and complex that only a skilled professional could install them, thus offering further returns to the neighbourhood electrical store, whereas distributive

The Great Hall of Stokesay Court, by Thomas Harris, 1889, with lights still arranged according to the original design.

solutions were often effected with table-lamps, standard-lamps and 'art lamps' generally, which the householder could plug in to a wall outlet without skilled assistance. The retail trade continued to combat the portable lamp for many decades, and as late as 1925 an editorial in *Lighting-fixtures and Lighting* was demanding rhetorically 'Are lamps forcing out Fixtures?'[15]

[15] *Lighting Fixtures and Lighting*, February 1925, p 24.

The same editorial also implied that the use of central fixtures was in the interest of the architect or interior designer, as well as the interest of the trade

> The public, ignorant on lighting ... leans towards lamps which, in many instances, are entirely inadequate, besides throwing decorative scheme out of balance ... Living room in a costly residence with thirteen lamps and without ceiling pieces or brackets, produces a frightful combination of colours—Yet some-one was to blame for the lighting scheme; was it architect, owner, or just indifference?[16]

[16] loc. cit.

Whatever the situation may have become by 1925, it seems possible that at the beginning of domestic electrification, architects had interpreted the situation in Slater's terms, and preferred distributive solutions. Thomas 'Victorian' Harris's installation in the great hall of Stokesay Court, in Shropshire, uses low-power bulbs dangling on flex from simple wooden brackets attached to each of the columns that support the upper gallery. The date of the design is 1889 and this is almost certainly the first house in England specifically designed for electric lighting, since lighting-wires can be seen installed in their present positions even in pictures of the house taken immediately after completion (in the photographic album now at the RIBA Library).[17]

[17] information from correspondence with the present Sir Philip Magnus, and Lady Magnus-Alford.

The best general overview of the art of electric lighting in its early stages in undoubtedly that afforded by the writings of Dr Louis Bell, above all, his *Art of Illumination* whose first edition appeared in 1902. At one point in his argument, Bell makes an ingenious objection to the fixity of fixtures, and to the unquestioning acceptance of novelty:

Professor Elihu Thompson once very shrewdly observed to the writer that if electric light had been in use for centuries and the candle had just been invented, it would be hailed as one of the great blessings of the century, on the ground that it is perfectly self-contained, always ready for use and perfectly mobile.

Now, gas and incandescents, while possessing many virtues, lack that of mobility. They are practically fixed where the builder or contractor found it most convenient to install them, for while tubes or wires can be led from fixtures to any points desired, these straggling adjuncts are sometimes out of order, often in the way, and always unsightly.[18]

[18] *The Art of Illumination*, New York, 2nd ed., 1912, p 208.

However, the candle business is clearly only a debating point, and for the rest of his argument, Bell speaks directly to the problem:

In domestic, as in other varieties of interior illumination, two courses are open to the designer. In the first place he can plan to have the whole space to be lighted brought uniformly, or with an approximation to uniformity, above a certain brilliancy, more or less approximating the effect of a room receiving daylight through its windows. Or, throwing aside any purpose to simulate daylight in intensity or distribution, he can put artificial light simply where it is needed, merely furnishing such a groundwork of general illumination as will serve the ends of art and convenience . . .

In electric lighting the most strenuous efforts are constantly being made to improve the efficiency of the incandescent lamps by a few per cent, and an assured gain of even ten per cent would be hailed with such a fanfare of advertising as has not been heard since the early days of the art. Yet, in lighting generally, and domestic lighting in particular, a little skill and tact in using the lights we now have, can effect an economy far greater than all the material improvements of the last twenty years. The fundamental rule of putting light only where it is most useful and concentrating it only where it is most needed, is one too often forgotten or unknown. If borne in mind, it not only reduces the cost of illumination, but improves its effect.[19]

[19] op. cit., pp 208–209.

But if these constitute a set of ground rules for the functional deployment of 'the lights we have' as of 1902, there were also visual and aesthetic problems of electric lighting that can be regarded as almost specific to the architect as artist. The use of electric lights could alter the appearance of forms and volumes in a way that no other environmental aid ever could. For instance, by concealing tubular strip lights (which existed from quite an early stage of the art) above projecting cornices, with their illumination thrown upward onto a vault above, it was possible to reverse the fall of light and shade across its curved surface, to invert normal visual expectations and make a nonsense of the ancient art of sciagraphy.

The possibilities and problems inherent in such employments of light (now commonly duplicated in exterior situations where floodlighting is employed for publicity or *Son et Lumière*) have barely been understood or usefully discussed in the eighty-odd years that electric lighting has been with us, though there is often implicit evidence that architects were aware of them. Another, and equally intriguing problem that arose does seem to have had some public discussion: in 1917, Morgan Brooks in some observations on *The Relation of Lighting to Architectural Interiors* wrote:

> At first sight it appears surprising that an architect who has success-fully produced a beautiful interior should relegate the lighting thereof to an uninspired subordinate, with results so inharmonious as to ob-scure his art. Doubtless, this is partly due to the fact that the architect did not visualise his illumination with his interior plan, and will not or cannot give it afterthought, and partly because he is not seriously dis-turbed by the incongruous lighting of an interior which appeals to him as beautiful with or without light, so powerful was his original idea.[20]

and the insight into the psychology of the architect, both as artist and professional, that he showed here does much to illuminate the whole problem of subsequent inabilities to grapple with the prob-lem of lighting. Architects are at the mercy of their first sketches, and those sketches normally represent forms viewed in natural day-light, or some form of abstract universal light such as only exists in architectural sketches. What it never is, or only very rarely, is light emanating from inside the illuminated objects, and therefore, as Brooks also observed:

> It has been customary enough for architects to design their specially-built gas and electric fixtures, but it will be agreed that, as a rule, the harmoniousness of these fixtures is felt more by day than by night.[21]

This is not only shrewd and a truism, but it also cuts very deep into the problem. How can a man trained to model forms by

[20] *Scientific American Supplement* (Vol. LXXXIII) June 2, 1917, p 367 (reprint of a paper given to the Illuminating Engineering Society).

[21] loc. cit.

external light and its cast shadows, to define architecture in Le Corbusier's terms as 'forms assembled in light', turn his art inside out and model his shapes by light emerging from within, and without shadows, to define his art as 'the magnificent, cunning and masterly play of light assembled in forms'?

Electric lighting thus put the challenge of environmental technology to architects in direct terms of the art of architecture, because the sheer abundance of light, in conjunction with large areas of transparent or translucent material effectively reversed all established visual habits by which buildings were seen. For the first time it was possible to conceive of buildings whose true nature could only be perceived after dark, when artificial light blazed out through their structure. And this possibility was realised and exploited without the support of any corpus of theory adapted to the new circumstances, or even of a workable vocabulary for describing these visual effects and their environmental consequences. No doubt this accounts for the numerous failures in this century to produce the effects and environments desired; equally doubtless it accounts for the periodic waves of revulsion against 'glass boxes' and fashionable returns to solid concrete and massive masonry, where visible form is still generated by external light and cast shadows, for which there is established theory and customary terminology.

We have been passing through such a period of revulsion and return in the last decade, and valid-sounding reasons can probably be advanced for it, such as the need to consolidate our knowledge and re-appraise our progress. But however it is excused, the fact remains that compared to the range of technological aids to environmental management currently available, the attitude of the architectural profession seems vastly less adventurous than that of the pace-setters of 1900–1914, and especially that of Frank Lloyd Wright who, by any standards, must be accounted the first master of the architecture of the well-tempered environment, and must therefore be the hero of the next two chapters.

5. The environments of large buildings

The kit of new mechanical devices for environmental management that existed by 1900 posed—it appeared—two different sets of radical problems/opportunities in architecture. One set had to do with changes in buildings that were *enforced* by the employment of new devices—especially finding room to accommodate the plant, and necessary structural changes such as improved insulation to extract reasonably economical performance from them. The other set had to do with changes in buildings *facilitated* by the new, devices, especially the freedoms accruing from not having to adapt the structure to husband, or create, particular environmental qualities. In practice, it is usually difficult to be certain which set of considerations were dominant, or how the two sets interacted, because in most of the buildings that are worth discussing submission and exploitation are inextricably entangled. Nevertheless, one can at least typify the architectural constraints imposed by environmental machinery from a study of larger buildings, and the benefits from a study of domestic structures, without doing too much violence to the historical record.

The very largeness of large buildings created new environmental problems, not only from the vaster bulks of structural material involved or the greater volumes of air enclosed, but also by upsetting exterior meteorological conditions by banking up wind-pressures, or overshadowing large areas of ground. Large mechanical devices were at hand to deal with at least the internal consequences of these disturbances of customary scale. But the sheer size of this machinery in its early states brought further problems in its wake. The difficulty of accommodating it anywhere but in the basement because of its weight has already been mentioned, but locating it there entrained other constraints. If it was steam-driven, for instance, it would need a fair height of chimney for its boiler-furnace. If, in addition, a massive extract duct was required, descending the full height of the building, plus an equally massive upcast exhaust duct to dispose of the extracted air, also rising the full height of the structure, then the design was lumbered from the start with three monumental vertical features, for which some accommodation had to be found inside or outside the building's perimeter, and which together might constitute an extremely expensive structural exercise. No wonder Plenum systems with natural exhaust held such attractions around 1900.

But it was not merely the sheer bulk of building to be ventilated, warmed or lighted that gave rise to unprecedented problems; their form and constructional techniques had environmental conse-quences too. Skyscraper office blocks in particular introduced novel discomforts and difficulties, which required urgent solution. Such matters normally receive scant treatment in the historical litera-ture, which commonly assumes that the steel frame and the elevator were all that were needed to make tall office blocks possible. In fact, as Burchard and Bush-Brown have rightly pointed out,[1] a gaggle of other devices, such as electric lighting and the telephone were equally necessary in order for business to proceed at all—and without ability for business to proceed, skyscrapers would never have happened. Yet even these authors do not mention the flush-ing W.C., for instance, without which such tower blocks would be uninhabitable, nor the various devices required to combat the thermal and ventilating peculiarities of the skyscraper as it had become established in Chicago and New York by 1900.

From any environmentalist's point of view, many of these skyscrapers were inherently unsatisfactory, and their short-comings were aggravated in practice by rising expectations of per-formance on the part of both users and building owners. Thus, Konrad Meier:

[1] John Ely Burchard and Albert Bush Brown, *The Architecture of America*, London, 2nd ed., 1967, p 157. The authors also have a certain amount to say, in general, about the consequences of electric lighting, but still come to the conventional conclusion that 'the most important of these develop-ments for the building art was the . . . use of steel' [p 156].

The standard of requirements has equally been raised, as to degree as well as to permissible variation in temperature . . . under structural conditions growing more and more adverse. Certain unpleasant experiences with some of the thin, tall and flimsy buildings now being erected, and not even suited to their purpose, will serve to illustrate this difficulty.[2]

[2] Meier, op. cit., p 4.

It is not common nowadays to think of the pioneer skeleton-frame skyscrapers as 'tall, thin and flimsy' and it is therefore sobering to reflect that famous works of architecture, such as Burnham and Root's Reliance Building, would fit comfortably into the class of buildings of which Meier complained. By comparison with the massive masonry structures of earlier decades, they were quite light enough to introduce novel 'unpleasant experiences,' though their deficiencies are less effectively summarised by Meier than by Bushnell and Orr in their textbook on district heating, where they refer to

. . . a skeleton or framework of steel columns and girders, enclosed by a brick wall and finished on the outside with brick or terra-cotta tile. With such tall buildings it is necessary to use the lightest material available in order to decrease the weight on the steelwork and foundations. In doing this, of course, the thinness of the walls becomes of importance from the standpoint of heating calculations. Such buildings have little capacity for storing or retaining heat, which is in contrast to what is found in buildings of massive masonry. In the former case— the modern building—heat must be furnished for a much longer daily period than the latter, due to the more rapid cooling effect. Furthermore, the modern structure is designed with a view to utilising as much of the exterior as possible for window-space, as by so doing the lighting conditions are vastly improved. In fact, some buildings are practically 40% to 45% glass area, and the heat loss from such buildings is proportionately high . . .

One peculiarity which has been noted sometimes in very high structures is the draft effect due to the inrush of cold air through openings on the lower floors. The air, on being heated rises rapidly through the various elevator and ventilation shafts, and causes a partial vacuum effect, with a useless expenditure for heating the large volume of air.[3]

[3] Bushnell and Orr, op. cit., p 207.

The nuisance value of this thermal siphon effect in tall buildings went beyond waste of heat however: the suction could pull in bad weather and street dirt at ground level, could make doors difficult to manage, and whisk papers from desks. The ultimate solution was to be controlled mechanical ventilation of a sealed building envelope, but a simple and ingenious Victorian solution was to hand by the end of the eighties—the revolving door. This was hardly a novel invention, but it was in this period that it was brought to its present level of operational perfection and was made a piece of standard equipment, specifiable ex catalogue, and it was for these refinements that Theophilus van Kannel received the John Scott medal of the Franklin Institute in Philadelphia in 1889. The van Kannel Company's slogan 'Always Closed' ('a person passing through the door pushes any one of its four wings forward, the wing behind him arriving at the curved side wall before the wing in front leaves it') explains well enough why their catalogue[4] could claim better ventilation control and greater uniformity of temperature within the building as a consequence of using their revolving door. It was an effective environmental filter that admitted persons but not the wind, a draught-lock if not an air lock, that strangled violent up-currents at birth.

[4] Van Kannell catalogue of 1901.

Such timely innovations, however, could not entirely deal with all the environmental problems of large buildings. Nor could innovations in their structural form do as much as might have been hoped, especially where constraints upon the plan were imposed by the nature of the site—that is to say, very little could be done to the architectural design of skyscrapers to improve their environmental performance while they stood tall upon such small sites; their faults derived from the economic and urbanistic situation that caused them to be built in a tall and narrow format. Radical mechanical improvements were to be the only solution, but it would be some time before mechanical plant had been sophisticated to the point where it could be installed in skyscrapers without cutting away so much rentable floor-space as to cancel any economic gain that might have accrued from environmental improvements.

The van Kannel revolving door unit in its most inexpensive and basic form, 1900.

On less constricted sites, innovations in the forms of large buildings could come some way to meet the environmental technology then available, and together they could offer significant improvements. Two outstanding examples may be cited from the first years of the present century, both motivated by an external climate containing a local excess of pollutants. In both of them architectural form and almost complete conscious control of the internal environmental conditions are inextricably entangled, but there resemblance ends—their architectures could not be more different. The less progressive in architectural style, but more advanced environmentally of the two, is the Royal Victoria Hospital in Belfast, Northern Ireland. At face-value, the credit for its design goes straightforwardly to the Birmingham architectural firm of Henman and Cooper, with Henry Lea as their engineering consultant, but a fog of rumour has always surrounded the design, because local pride insists that the whole concept is too original to chime with the rest of Henman and Cooper's work, and at first sight it must appear strange that in a city where the forced ventilation of ships was a technological habit, and where Samuel Cleland Davidson's Sirocco works was producing some of the world's most advanced centrifugal fans, neither influence should have had any apparent direct effect upon the design. Davidson's, it is clear, were responsible for the design, installation and subsequent maintenance of the heating and ventilation machinery, and if their influence on the architectural concept was not direct, it could still have operated more obliquely through Davidson's own business and social connections. Shipping and ship-building interests were strongly represented on the hospital's board of management (as on most other things in Belfast) and the suspicion that some of them may have talked the architects into the final bulk form of the building is heightened by thinly veiled accusations[5] that the discussion at the Royal Institute of British Architects which followed the presentation of the design by the architects, was rigged and deliberately talked out of time by the scheme's supporters in order to prevent awkward questions being asked.

[5] correspondence in *The Building News* (and elsewhere) 1903–1905. The original presentation of the building (still a fundamental document) and the subsequent discussion at the RIBA are reprinted in *RIBA Journal*, Vol. XI, 3rd series, 1903–1904, pp 89*ff*.

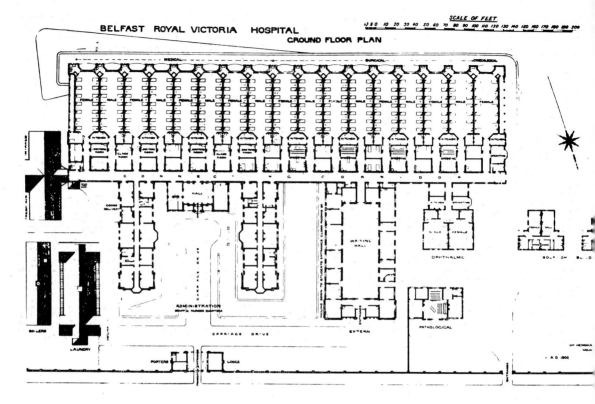

BELFAST ROYAL VICTORIA HOSPITAL
GROUND FLOOR PLAN

SCALE OF FEET

Be all this as it may, there can be no doubt about the propriety of making a public presentation of the design in the architectural profession's highest forum in Britain, because it represents a level of mechanical innovation and originality of plan that would have been hard to equal anywhere at the time. Within the work of Henman and Cooper themselves, the progress of innovation is clear enough. In their design for the Birmingham General Hospital in 1893, they had applied William Key's Plenum system of ventilation to a hospital organised on the traditional pavilion plan. This was an illogical and wasteful solution because the whole hygienic motivation of the use of separate pavilions had been to promote

Plan at main floor level of the Royal Victoria Hospital, Belfast, 1903, by Henman and Cooper.

good natural ventilation between windows and other openings on opposite sides of relatively tall narrow buildings, such as survive in many British hospitals from the last century.[6] With forced ventilation of any sort, this type of planning became unnecessary, and the large amount of external surface relative to the floor area would lead to a relatively uneconomic waste of heat. To get full advantage from the Plenum system, a much more compact plan would obviously be required, with as much as possible of the accommodation packed into a single structural volume (in order to avoid heat wastage from underground ducts between separated buildings).

These desiderata were fully achieved in the Royal Victoria Hospital, Belfast, where the maximum compactness of plan is

Exposed end of the ward block, Royal Victoria Hospital, as originally completed, showing ventilating turrets.

[6] on the origins of the pavilion plan as a ventilating solution in hospital design, see; Anthony King, *Hospital Planning: Revised Thoughts on the Origin of the Pavilion Principle in England* (Medical History, Vol. X, No. 4, October 1966).

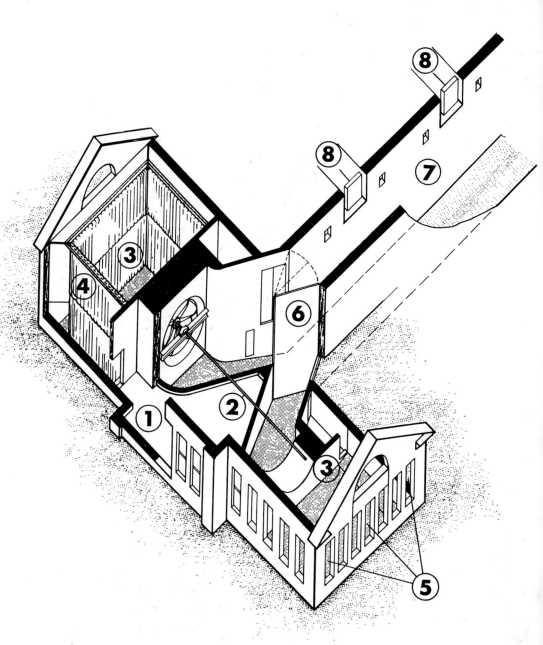

Left: cutaway section of engine house and head of main duct in the Royal Victoria Hospital, Belfast.

1. Engine room
2. Fan shaft
3. Heating chamber
4. Filtering ropes
5. Air inlet grilles
6. Draught control door
7. Main duct
8. Branch ducts

Right: cutaway of the complete ventilating system.

1. Fan house
2. Main duct
3. Branch ducts
4. Pipe runs
5. Air inlets to wards
6. Extracts from wards
7. Foul air extract duct
8. Foul air exhaust
9. Ward roof
10. Roof of operating theatres, etc.
11. Roof of main corridor

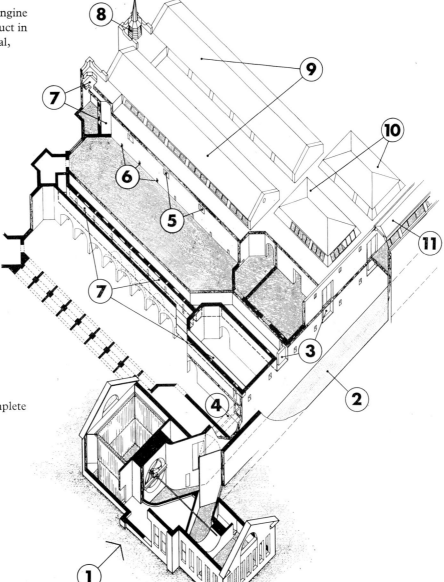

79

combined with a minimum length of duct. Nevertheless, the duct is one of the most monumental in the history of environmental engineering; a brick tunnel with a concrete floor, over five hundred feet long and nine feet wide, twenty feet deep at the input end, tapering upwards to only six feet deep at the downstream end. As may be deduced, this was no high-velocity system—a set of steam engines operated by waste steam from the hospital laundry's boilers drove a pair of slow-turning axial flow fans on a common shaft in an engine house at the input end of the duct. The output from these fans fed together into a Y-branched inlet, and thus into the duct, up which warm air moved at a little over walking pace in winter, and cool air somewhat faster in summer.

From the giant duct, the air was fed into distributor channels (also of brick and concrete) opening off the left-hand wall of the duct and high under its ceiling. From the distributors, the air rose through risers in the walls of the wards and was delivered through openings above head level into the wards on either side. For the wards were laid out in a tight parallel plan, with only party walls between them, single storey, and served by a common corridor the same width as the main duct and running its full length. Since the wards were butted up solidly edge to edge, and the intervening walls were not available for windows, the wards were lit by long lay-lights on either side of their pitched roofs, and by arched openings onto the balconies at the end. At the corridor end, kitchens, operating theatres, private wards, were fitted between the entrances to the wards. Thus the whole medical work space of the hospital was packed into a densely occupied single-storey block, divided into rooms, all top lit, that received tempered air from registers above head level, and disposed of foul air through slots in the skirting around the perimeter of the rooms, whence it descended into extract ducts parallel to the input distributors, and finally left the building through vertical risers terminating in the louvered lanterns, one at the end of each ward, which are a conspicuous feature of all external views of the building.

The degree of control designed by Davidson into the air-supply system was rigorous by the standards of the time (1903) and made ingenious use of the comparatively crude technology available. Air, on entering through window-type openings in the ends of the engine-house, was pulled through hanging curtains of coconut fibre robes kept moist by sprinklers in the roof of the filter-chamber (this water could be warmed in winter to prevent freezing). Cleaned of soot, smuts and other impurities for which the Belfast atmosphere was notorious, the air then passed through batteries of heating-coils before entering the fans and being propelled up the duct. In order to prevent excessive temperature drop during its relatively leisurely journey to the far end of the duct, the air received extra heat in winter from a booster pipe running its full length and from further coils of pipe in the entrance to each distributor.

This system, intended first to provide warmed and cleaned air to all the medical and surgical areas of the hospital, brought with it an additional benefit. Because the outside air was dirtiest in winter, that was also the season when the sprinkler system to moisten the filters was most used. But the winter was also the time when there was the biggest difference in external and internal air temperatures—air entering the system below freezing would leave it at a temperature in the sixties Fahrenheit—and thus the greatest reduction in relative humidity, in any system that did not make good the deficiency in the water-vapour content of the air. But this deficiency was, crudely, made good by the sprinkler system, more or less in direct ratio to need, because the colder the day, the more soot would be pumped into the Belfast atmosphere, and the more water would be run through the sprinklers and taken up by the air that passed through the ropes.

Had this topping-up of the relative humidity of the atmosphere been left to happy accident, there would be no point in enquiring whether the Royal Victoria Hospital has a place among pioneer air conditioning systems, but it was never left entirely to accident.

From the beginning, Davidson had intended some form of humidity control, and before 1920, if medical memories are to be trusted, it became the practice to observe the humidity of the air in the wards night and morning, by the normal wet-bulb/dry-bulb thermometer technique, and instruct the engineman to regulate the flow of water over the filter ropes accordingly. This immediately puts the historical argument on a different footing, for as soon as the humidity of the air is consciously regulated, along with the temperature and cleanliness, then there was air-conditioning in the strict sense laid down by Stuart Cramer in his fundamental patents of 1906

> I have used the term 'air-conditioning' to include humidifying and air cleaning and heating and ventilation.[7]

or as emphasised much later by Willis Carrier

> ... added to the control of humidity are the control of temperature by either heating or cooling, purification of the air by washing or filtering the air, and the control of air-motion and ventilation.[8]

But, since these conditions were achieved before the conscious art of conditioning air had been even given a name, and may not have become fully deliberate until after the art had become conscious, the position of the Royal Victoria Hospital among the pioneers of air-conditioning remains a little difficult to assess without becoming involved in semantic niceties. On the other hand it does seem true to say that it must have been the first major building to be air-conditioned for human comfort—earlier installations were exclusively industrial, and the deliberate air-conditioning of cinemas in the US came only in 1922, which seems definitely to be later than the beginning of conscious and continuous humidity control at the Royal Victoria Hospital.

In any case, the importance of the hospital in the history of architecture lies in its total adaptation in section and plan to the environmental system employed. What makes it even more interesting historically is that more than one environmental system is

[7] lecture given in 1906 and cited by Ingels.

[8] statement of 1949, also cited by Ingels.

employed, the architecture changing to suit, for a great part of the accommodation to the 'right' of the great corridor and duct was not served by the conditioning system, which was reserved for areas occupied by the sick. The areas occupied by the fit—office-workers, nurses off-duty, etc.—had conventional heating by gas fires and conventional ventilation by means of opening windows, since it was known that the fit would insist on controlling their own environment, even if the sick could not.

What is immediately striking about these areas of conventional environmental control is that they also revert to conventional architectural form in plan and section; they are relatively tall (five full storeys) and thin, so that they both catch the wind, and allow it to pass through any random combination of opened windows on either side of the block. They are, in fact, close relations of the pavilioned ward blocks that Henman and Cooper, and the rest of the profession, had become accustomed to design. The external massing of the various parts of the hospital thus gives direct 'expression' to two different kinds of environmental management, a low, top-lit format corresponding to mechanical systems, and a tall, side-lit format to natural systems.

The external aspect of the Royal Victoria Hospital also demonstrates with painful clarity the total irrelevance of detailed architectural 'style' to the modernity of the functional and environmental parts. As will now be clear, the hospital is extremely 'modern' and ahead of its time in its environmental controls; and it is also very modern in the way the parts are functionally disposed along a spine corridor without regard for axial symmetry—in these aspects of plan and circulation it approximates to the advanced practices of some thirty years later, and in the implied extensibility of its plan along the line of the corridor, it is still of interest to proponents of 'indeterminate' architecture some sixty years later. But in its detailing, in what its designers doubtless regarded as its 'art architecture' it belongs dismally and irrevocably to a conception of 'Welfare' architecture fathered by the London School

Board some forty years before, a style already thoroughly discounted and out of fashion among consciously progressive architects of 1900.

Without doubt it is this depressingly un-modish exterior that has caused the Royal Victoria Hospital to be universally overlooked by historians of the Modern Movement until the last few years. It is not mentioned by any of the historian-protagonists of Functionalism; it receives not even a footnote in Pevsner's *Pioneers of the Modern Movement*, in spite of the fact that in all except the purely stylistic sense it was far more modern and far more pioneering than anything that had been designed by Walter Gropius, the hero-figure of the book, before 1914.

But then, Pevsner also fails to draw attention to the environmental innovations in one of the buildings that he does emphatically find worthy of mention. Yet the School of Art in Glasgow is not only a near contemporary of the Royal Victoria Hospital, but also used a Plenum ventilation system—which is not surprising in William Key's home town—whose upcast ducts appear, uncommented, in practically all the standard photographs of the studios and work-spaces of the school. The provision of such a system of hot air ventilation and heating was a necessary concomitant of Charles Rennie Mackintosh's use of huge north-facing windows in these rooms, and a humane provision where the life-class is concerned, for Glasgow is a chill city for nude models. Furthermore, the abandonment of the Plenum system (due to insufficient upkeep rather than inherent faults in the design) and its substitution by an extended hot-water radiator system has done much to destroy the spatial artistry for which Mackintosh is renowned and of which the school is reckoned his masterpiece. In the celebrated library, for instance, the tall oriel window-bays are now almost impossible of access because they are fenced off by radiators insensitively placed across their openings.

These points are raised not out of hostility to Pevsner (who has professed himself 'happy' and 'puzzled' to claim the present

Above: ventilating grille in duct in a studio, Glasgow School of Art.

Right: part plan of the Glasgow School of Art, 1904, by Charles Rennie Mackintosh, showing ventilating ducts on central walls.

R = radiator
S = sink
ARROWS indicate input and extract ducts

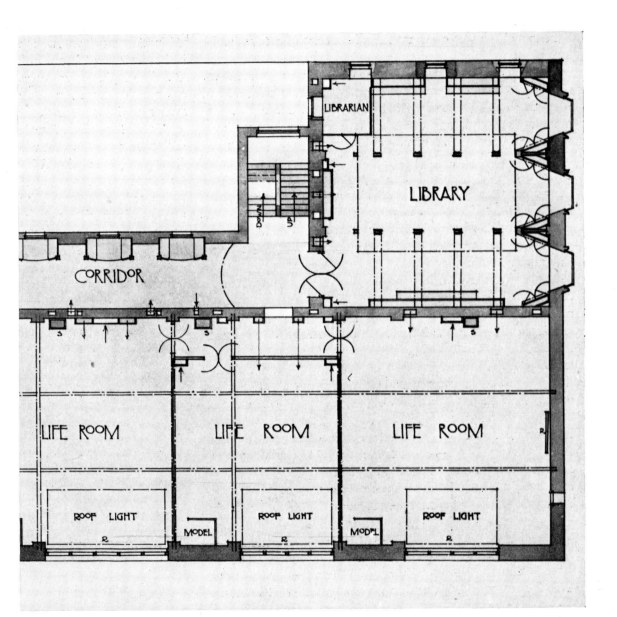

author as his pupil) but as a complaint against the general design-blindness of the whole generation of historians of modern architure whose writings helped establish the canons of modernity and architectural greatness in the present century. All would insist, and rightly, that new concepts of space, such as were pioneered by Mackintosh at the Glasgow Art School, are crucial to their conception of modernity, none seem to have acknowledged that such spaces would usually be uninhabitable (and therefore not architecture) without massive contributions from the arts of mechanical environment-management. Narrowly pre-occupied with innovations in the arts of structure, they seem never to have observed that free-flowing interior spaces and open plans, as well as the visual interpenetration of indoor and outer space by way of vast areas of glass, all pre-suppose considerable expense of thermal power and/or air-control, at the very least. These lacks in historiographical method appear very clearly in the literature about the only building contemporary with the Royal Victoria Hospital whose environmental architecture can really be compared with it in radicalism and ingenuity—Frank Lloyd Wright's Larkin Office building in Buffalo, New York. No-one doubts that it was one of the early masterpieces of pioneer, modern architecture, and if Wright had designed nothing afterwards he would still have held a strong place among the fathers of twentieth-century design. Because its exterior was expertly pleasing and conformed to the stripped classical taste favoured by the other pioneers, especially the Europeans like Perret, Garnier, Behrens and Gropius, the Larkin building finds a natural place in the history books, unlike the Royal Victoria Hospital. But like the hospital, it has a neatly equivocal place in the environmental history of its times, for, as Wright himself put it:

Facing page: Larkin Administration Building, Buffalo, N.Y., 1906, by Frank Lloyd Wright; above: exterior showing stair-towers and duct-boxes at corners; below: one of the working galleries, with ventilating grilles visible under edge-beam of balcony above.

> The Larkin administration building was a simple cliff of brick hermetically sealed (one of the first 'air-conditioned' buildings in the country) to keep the interior space clear of the poisonous gases in the smoke from the New York Central trains that puffed along beside it. [9]

This is a strong claim; most historians have either passed it over

[9] Frank Lloyd Wright, *An Autobiography*, New York, 1943, p 150.

86

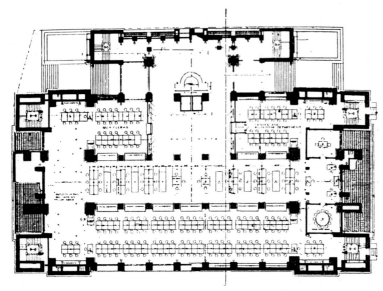

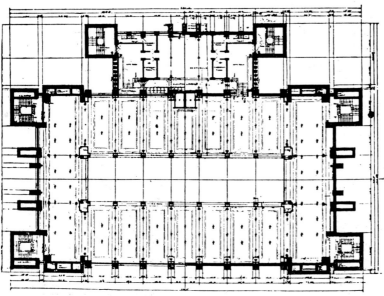

Larkin Building; top: plan of entrance level and, bottom, of typical working floor.

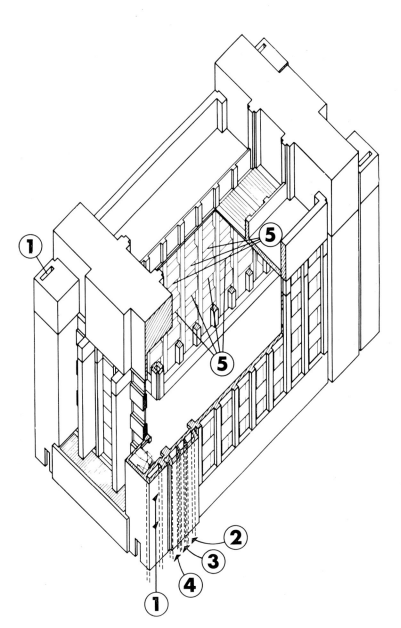

Larkin Building: cut-away drawing showing location of main air-ducts.

1. Fresh air intake
2. Tempered air distribution
3. Foul air and exhaust
4. Utilities duct
5. Tempered air outlet grilles under edge of balconies

as if unnoticed, or—what is worse—repeated it without the quotation marks which Wright knowingly put round the words 'air-conditioned' (on which more, below). Historical and critical writing has tended to concentrate exclusively on the felicity of its interior spaces and their relationship to the great monumental volumes of the exterior, without observing that the system of environmental management mediates crucially between interior and exterior form. Not only is the use of a vast single vessel of space ringed with balconies almost a necessity given the then state of artificial ventilation, but it was a flash of inspiration, about the disposition of the services to achieve that ventilation, that gave the magisterial form of the exterior. Wright again:

> But not until the contract had been let to Paul Mueller and the plaster model of the building stood completed on the big detail board at the centre of the Oak Park draughting room, did I get the articulation I finally wanted. The solution that had hung fire came in a flash. I took the next train to Buffalo to try and get the Larkin Company to see that it was worth thirty thousand dollars more to build the stair-towers free of the central block, not only as an independent stair-towers for communication and escape, but also as air-intakes for the ventilating system.[10]

[10] loc. cit.

The Larkin Building no longer exists, having been demolished in 1950, but the surviving paper evidence—including a few photographs taken during demolition—serve to confirm how crucial was that flash of inspiration about where to put the services, in giving the elusive articulation that was afterwards to make the design so famous. As is well-known, the basic *parti* of Wright's layout surrounds the galleried central space with four corner towers, and flanking entrance unit, lying on the transverse axis to one side. Much of the inner wall surface of the working galleries, including the backs of the solid balustrades, was given over to modular filing cabinets, and the external windows, closed against the unwelcome atmosphere outside, were set high under the ceilings of each tier, above the filing cabinets.

The internal atmosphere was serviced as follows: air from well above the external pollutants was drawn down capacious ducts in the blank walls of the corner towers, at the sides of the staircases, as Wright indicates. In the basement it was cleaned and heated, or after the installation of the Kroeschell refrigerating plant in 1909, cooled[11]—but never humidity-controlled, and hence Wright's judicious quotation-marks around the words 'air-conditioned' (in the town where Carrier was perfecting humidity control he had better be careful!). The tempered air was then blown up riser ducts in the massive blank brick panels on the exterior walls immediately adjacent to the stair-towers, and distributed floor by floor through input registers on the backs of the downstand beams under the balustrades of the balconies. The same blank brick panels also contain the exhaust ducts through which vitiated air is extracted, and a third duct-space in the panel houses pipes and wiring and other ancillaries. Throughout the interior, extracts are marked by a characteristic and much-used Wrightian device—a pattern of hollow bricks forming a coarse grille, around which even the earliest photographs show the typical staining that commonly marks an exhaust.

[11] date given in Ingels' Chronology.

Neat as these usages may be, and monumental as their external consequences may be, the Larkin Building, like the Royal Victoria Hospital, must be judged a design whose final form was imposed by the method of environmental management employed, rather than one whose form derived from the exploitation of an environmental method. This is in no way to denigrate the masterly manner in which Wright managed to turn those impositions to his architectural purposes. Indeed, that flash of inspiration to which he attached such importance lay largely in the sudden vision that purely architectural ideas contained in the design at the stage of the plaster model, also contained the seeds of a solution that would both deal handsomely with the services, and produce an even more striking exterior—the published drawings of the earlier version already have in embryo, and shallow relief, the forms that were to appear in full 'articulation' in the final version.

In this, the Larkin building is something of a watershed. The Royal Victoria Hospital and its innumerable predecessors in the realm of 'welfare architecture' for industrialised mass-societies, often present the image of new functional needs and mechanical possibilities bursting through a crust of conventionally conceived architectural forms, whereas in the Larkin Building, the transformation of the exterior, both in detail and in mass, appears to be keeping pace with transformation of the interior economy of the building-type. Thus Wright, in the Larkin Building design, serves as a bridge between the history of modern architecture as commonly written—the progress of structure and external form—and a history of modern architecture understood as the progress of creating human environments.

And it is important at this point to insist upon the totality of the human environment, and not to become mesmerised by the technological innovations that went into its creation, for the houses that Wright was designing in the same years established equally radical improvements in the art of environment without embracing any technological devices that were spectacularly novel in 1900. Just as their visual style remained unpretentiously picturesque in most cases, and has nothing to do with the Machine Aesthetic which so often claimed Wright among its ancestors, so their level of mechanical equipment was usually modest, and the nature of that equipment, well-tried and familiar. It was the use he made of mechanics and structural form in combination that marks out the Prairie Houses as triumphs of environmental art.

6. The well-tempered home

Many of the virtues of Frank Lloyd Wright's domestic designs have always been immediately apparent to all who study architecture. But to do any kind of justice to their environmental qualities, they must be seen against the background of the domestic work of Wright's contemporaries, both in America and Europe, and of the American domestic vernacular on which both Wright and his American contemporaries were able to build.

The peculiar status of that generation of US domestic designers whom we are accustomed to group, loosely, as 'Wright and his California Contemporaries' has been increasingly noticed since Henry-Russell Hitchcock drew them together (in defiance of previous historiographical custom) in a single chapter of *Architecture: Nineteenth and Twentieth Centuries*.[1] At a purely visual level their work before 1914 is unified by conspicuous spatial innovations lurking behind an apparently dis-unifying variety of romantic stylings—Japanese, Tudor, Mission-style, Italianate, etc. This superficial variety does indicate an important underlying sociological unity, since it demonstrates that the architects were working *with* the artistic preferences of their upper-middle class suburban patrons, and not—as was so often the case in the less relaxed intellectual atmosphere of Europe—*against* them. A plain confrontation of houses by Adolf Loos and those of even so restrained a California architect of his time as Irving Gill, will show that, in spite of startling occasional similarities of detail and undecorated exterior form, the cultural pre-suppositions underlying them must be poles apart. These diametrical oppositions of intention were perceptively noted by J. J. P. Oud, in Holland, in his observation on Wright:

[1] Harmondsworth, 1958.

That which in Cubism—and it cannot be otherwise—is puritanical asceticism, spiritual self-denial, is with Wright exuberant plasticity, sensuous superfluity. What arises in Wright from the fulness of life to a degree that could only fit into an American 'High-life', withdraws in Europe to an abstraction that derives from other ideals and embraces all men and everything.[2]

[2] J. J. P. Oud, *Hollandische Architektur* (Bauhausbuch 10), Jena, 1926, p 81.

The distinction is well made, and upon the right grounds, but the conclusion Oud draws will not stand. The abstract European style of which he speaks was no more than an ideal, and far from embracing all things and all men, embraced only the upper strata of the European intelligentsia, and that for a limited period only, even though its proponents called it the International Style, and its historians prematurely assumed it to be the great style of the twentieth century. On the other hand, versions of the kind of architecture promoted by Wright and his California contemporaries are now much more nearly universal and international than the derivatives of the old International style.

For this situation, the adoption of some aspects of the Wright/ California idiom as the international norm for hotels and motels is doubtless largely responsible, even if the idiom has been thinned out till little remains but wood-graining and wide horizontal shelves and rough-textured textiles. But this idiom has genuine virtues—it is visually undemanding, acoustically quiet, thermally comfortable because its vegetable-fibre surfaces are quick to warm. Furthermore, these surfaces and forms seemed to settle well with central heating, electric lighting or (later) air-conditioning, and they had so settled long before the style began to spread beyond the Chicago and California suburbs that gave it birth. The idiom and the technological supports grew together from an unified approach in the minds of the architects. Alan Gowans (in *Images of American Living*) identifies an 'interest in human engineering' as the unifying factor, and observes that given this interest

It is no accident . . . that Wright's Larkin Building introduced such conveniences as an early air-conditioning system (*sic*), new designs in metal office furniture and lighting fixtures. Or that Gill was famous for developing coloured and waxed concrete, garbage disposals, vacuum cleaner outlets, automatic car-washing devices in his garages and what he described generally as 'the idea of producing a perfectly sanitary, labour-saving house, one where the maximum of comfort may be had with the minimum of drudgery.' Or that Maybeck should invent burlap sacking coated with foamy 'bubble stone' concrete for fireproofing houses in areas prone to bush-fires, built in ovens . . .[3]

[3] Alan Gowans, *Images of American Living*, New York, 1964, p 407.

But Gowans is, apparently, surprised to find a disunity between the environmental consequences of this interest in human engineering and the external style of some of the buildings in which it was put to work, even though he finds it easy to

. . . explain, in consequence, why, despite the fall from favour of the Prairie architects and the California School generally after 1910, their work continued to have such a hidden, but for all that the more pro-profound, influence on the suburban period house in the following two decades.[4]

[4] loc. cit.

What needs to be added to Gowans' observation is the fact that the external styling of such houses had already become irrelevant even in the years before the two decades of the 'suburban period house,' and that the usefulness and applicability of the interest in human engineering lay, partly, in the fact that it had been detached, effectively, from any architectural style as normally conceived. The Wright-California 'idiom' discussed above is a way of organising interiors, and derives from a shift of emphasis from exterior show in domestic architecture to interior comfort in domestic environment. That emphasis could be packaged in any number of ways, largely because the package was becoming rather irrelevant. If this statement be doubted, in view of the overwhelming visual importance of roofs in both the Prairie and California styles, then it becomes pertinent to ask what it was that the roofs had overwhelmed. The answer is, of course, the walls, normally the most conspicuous and elaborate parts of the package, the surfaces on which architectural pretensions had been most assertively inscribed. In European modern house-architecture, especially after Adolf Loos, the walls become the master elements and roofs were banished from view, but in the US, even those who have much to say about walls, like Frank Lloyd Wright, are proud to have broken out of them.

This development was not initiated by Wright and his California contemporaries; they exploited to full design-advantage a tendency which is probably as old as North American domestic building of any sort, a tendency whose conscious exploitation can already be dimly perceived in the writings of nineteenth-century domestic reformers, such as the redoubtable Catherine Beecher. James Marston Fitch, having rightly singled her out as one of the most important foremothers of the US suburban way of life, draws attention to the difference in the house-types she offers in her *Domestic Economy* of 1842, and her *American Woman's Home* of 1869. The first has a conspicuously neo-Grec exterior, complete with pilaster strips, and chimneys up the outer walls which register as a species of acroteria on the elevation, and an interior conventionally carved into boxy rooms. The whole house could equally well find itself in Boston, Park Village in London, or even Berlin, at that time, without evoking surprise or even interest.

The house of 1869, however, the fruit of Catherine Beecher's exposures to the facts of life and technology in the newly opened middle-west, represents a total transformation of her conception of domestic design. Modest gothic detailing on the exterior does not disguise the fact that half the elevation is now roof, but does nothing to explicate the interior planning, which is the most consequential of the innovations reflected by this design. It seems to introduce for the first time the conception of an unified central core of services, around which the floors of the house are deployed less as agglomerations of rooms, than as free space, open in layout but differentiated functionally by specialised built-in furniture and equipment, thus anticipating the basic functional organisation of Buckminster Fuller's Dymaxion house of 1927. These innovations are so radical and so original that Fitch has no compunction in drawing parallels between the American Woman's Home and the

European modern architecture of the 1920's:

> ... the house itself—while still nominally 'Gothic' in style—is now firmly visualised as a true machine for living. No longer are there generalised or anonymous spaces; from top to bottom every cubic inch has been carefully analysed and organised for a specific purpose. Classified storage for all household objects is now fully realised as the first step in efficient house-keeping. In the kitchen it produces cabinet-work of astonishing modernity with shelving, cupboards, drawers and countertops which anticipate current practice.[5]

[5] *Architecture and the Esthetics ... etc.,* p 77.

But if this sounds like European rationalism of the period of le Corbusier's *machine à habiter*, it also sounds like the 'current practice' in kitchen-furniture design revealed by the catalogues of the great US mail-order companies. The sundry Colonial, and other, styles of that furniture suggest that its current practices might have been much the same had European rationalism never flourished. Furthermore, the intensity and rigour of the classified storage provided for domestic objects, and the sheer quantity of it provided, require some qualifications to Fitch's proposition that there are no 'general or anonymous spaces.' This is true in the sense that there are no cupboards or rooms offered simply as bulk storage for any objects or services that might be crammed into them, but the fact that there is now a specific place for every separate and identifiable *thing* leaves the rest of the space much more at the householder's disposal. For instance, it becomes much more simple and practicable to convert a living room into a temporary bedroom for a visiting preacher, because part of Catherine Beecher's classified storage is in cupboards on rollers which can be moved around to form screens or room dividers.

Furthermore, the manner of concentrating the household services tends to free much of the surrounding space from fixed functions dictated by specific pieces of service equipment. The heating provisions, for instance, appear intended to make all parts of the house equally usable. It is not merely their sophistication—

> Her services are quite complex and highly developed. She links a base-ment hot-air furnace, Franklin stoves and a kitchen range into a central

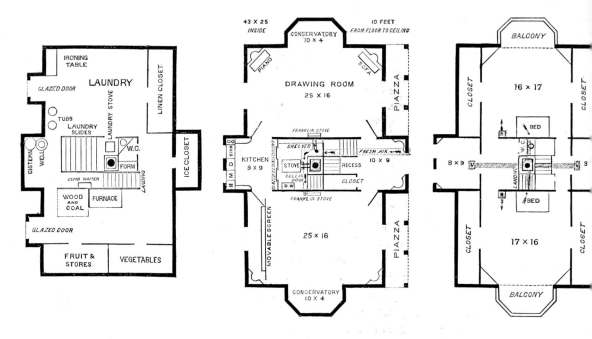

American Woman's Home (project), 1869, Catherine Beecher; above, left to right: plans of basement, ground floor and bedroom floor.

heating and ventilating system of some sophistication. She has eliminated all fireplaces as dirty and inefficient. The house is now served with an essentially modern plumbing system . . .[6]

but that the environmental servicing of the whole house has—at least, in intention—been consciously taken in hand by someone who believed herself (with some justification) up-to-date with the latest knowledge in the field. Draughts are killed at source by under-floor ducting feeding fresh air to the stoves, warmed air is delivered to the points where it is required, and foul air is extracted from the points where it is not.

All this is meant to be achieved by a single heated flue and extract, around which the equipment clusters, sending out good air and hot water, and into which foul air is recalled and disposed of.

[6] loc. cit.

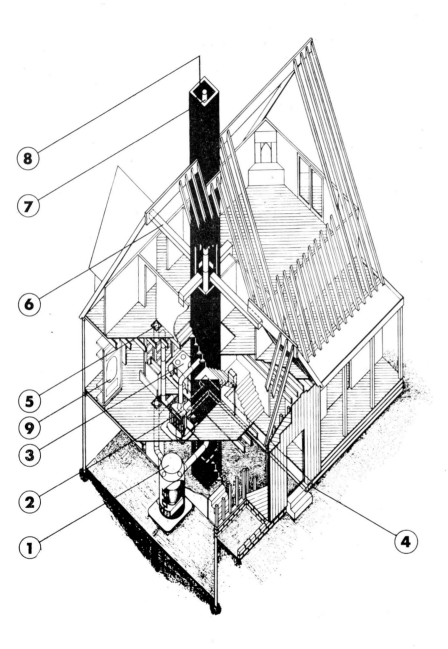

American Woman's
Home: cut-away show-
ing the complete house
as an environmental
system.

1. Hot air stove
2. Franklin stove
3. Cooking range
4. Fresh air intake
5. Hot air outlet
6. Foul air extracts
7. Central flue
8. Foul air chimney
9. Movable wardrobe

The point about this environmental 'tree' and its branching ducts is that it has robbed the outer wall of the house of all traditional environmental functions bar two—keeping weather out and letting light in. It carries no fireplaces or chimneys, no water-piping of any consequences, nor—since one may safely posit a lightweight, balloon-frame construction by this date—does it act as much of a thermal barrier. It has become, as Groff Conklin was to phrase it much later

> . . . nothing but a hollow shell . . . And most shells in nature are extraordinarily inefficient barriers to cold or heat.[7]

[7] *The Weather-conditioned House*, New York, 1950, p 15.

With little left to do, except to prevent the tempered atmosphere within from blowing away, the outer shell of the American house, both in Catherine Beecher's early vision, and in later built fact from coast to coast, lost most of its detailed relationship to the internal economy and layout of the house, and thus became susceptible of unrestrained stylistic diversification according to any fancy or inhibition that came along.

This independence of the outer shell is probably latent in the domestic technology of all wood-building cultures. Those of northern Europe, including Norway and much of Russia, usually possess a primitive vernacular plan-type in which an open hearth or massive stove is located in the centre of the house. In Norwegian usage, even when the smoke-stove was provided with a chimney and moved, for convenience, to the perimeter of the house-plan, it still stood within the skin of the wooden wall, with a clear space between the masonry of the hearth and the woodwork. Later still, architecturally sophisticated buildings in wood in Norway, such as the Stiftsgården palace in Trondheim, move the fireplaces back to the centre of the thickness of the building, where they are backed up in pairs on the internal partition walls and the chimneys emerge through the ridge of the roof (as does that of Catherine Beecher's proposal).

But in the wood-building traditions of New England, the brick

Traditional Norwegian farmhouse with central fire-place (after Kavli).

chimneys were frequently placed on the exterior of the building, outside the wooden skin. This is manifestly absurd, since it involves almost a maximum waste of heat as well as numerous difficulties and inconveniences of construction and use, and must presumably have been a residual habit remaining from European brick building traditions, or from the structures so built that the early colonists took as models for their consciously 'representational' architecture in court-houses and the like. As buildings became lighter in construction with the introduction of balloon frame techniques after the 1840's, there was an increasing need to deploy heat swiftly and directly through the interior space. The iron stove, Franklin or otherwise, made it possible to bring the heat source as far into the interior of the house as one cared to run the horizontal part of the flue from the outer wall, but the logical solution to the employment of stoves was, clearly, to bring them to the centre of the house and use a central flue as Catherine Beecher proposed.

Other methods of heating and ventilating have rendered her technology obsolete in detail, but this house type she proposed, with a thermal performance and thermal needs resembling those of the flimsy skyscrapers discussed in the previous chapter, is in all environmental and most structural essentials the house that most Americans inhabit and most American tract-developers are building, a clear century after her book was published. Its innumerable advantages and manifest suitability to the way of life its inhabitants appear to prefer, have to be set against one inherent environmental defect of some gravity—its inability to deal with the heat and humidity of the summers to which most of the Continental USA is subjected. The lightweight shell admits heat all too readily, the compact plan does little to encourage ventilation or the dissipation of kitchen heat, so that the indoor micro-climate of the bedrooms under the pitch of the roof can hit one in the face on ascending the stairs like the blast from an opened furnace door.

Since 1950, of course, this problem has been fairly easy to

The Gamble House, Pasadena, Cal., 1908, by Charles and Henry Greene; sleeping porches on garden elevations.

solve with the packaged domestic air-conditioner; at any time before 1950 to make such a house habitable during the heat of summer was a triumph of good design and good management, and almost certainly involved sacrifices in other directions. The first, and almost the last to achieve this triumph were the architects of the Prairie and California schools, and they achieved it at the cost only of some mild prodigality of structure and floor-space (which often brought other countervailing benefits) and some prodigality of winter heating in Chicago.

In California, where winter heating is a more marginal problem, a range of structural solutions to the summer heat problem was possible, and the great names of the school there devised, between them, a graded spectrum of methods. Irving Gill, with his relatively massive concrete walls offering analogies to the environmental performance of traditional adobe construction, nevertheless made extensive use on occasions of shaded external walks and galleries. Bernard Maybeck, more prodigal in every sense with structure, came to make ever more use of overhanging roofs and pergolas in his later domestic work, after 1913. In this he may well have been influenced by Greene and Greene, whose prodigality

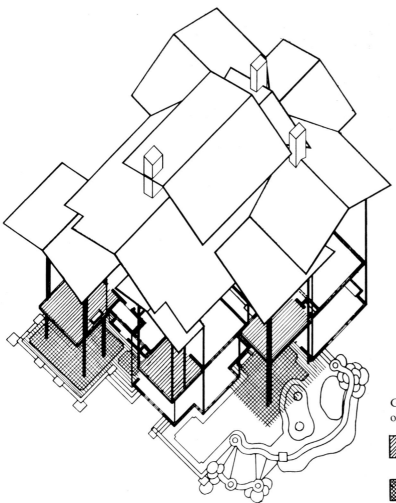

Gamble House; diagram of system of roofs, porches and terraces.

 Shaded external areas of sleeping porches

 Shaded areas of terrace at ground level

of roof-structure once reached the point where it seemed that their architecture was all roof.

That point is, notoriously, the summer 'cottage' they built for the Gamble family at Pasadena in 1908. Here, the widely projecting roofs over most gables are joined by an elaborate system of external roofed sleeping galleries on the upper floor and terraces at ground floor level, until external covered floor-space is almost equal to the floor-space inside the walls. The oriental inspiration of the detailing of the house is patent, but the scale of the conception and its opulence go beyond any conceivable oriental models. And all this, of course, to take advantage of the light southern California breezes and to shelter the walls from direct solar heating. In winter, to stop the same breezes wafting away the small amount of heating required (a conventional hot-air stove in the basement feeding a few registers in the floor, supplementing some largely ceremonial open fire places in the main rooms) the light walls were perfectly sufficient.

The purely structural aspects of the solution offered in Pasadena by the Greene brothers are very similar to those which Frank Lloyd Wright had evolved by the same years in Chicago. But whereas massive overhangs and cunningly placed openings will control the worst effects of summer heat by providing shade and exploiting through breezes, they do little for the inhabitants of the house during the deep cold of winter except, perhaps, to reduce damage to the structure by discouraging snow from drifting against the walls. It is quite likely that Wright, impelled by a climate that, from winter to summer, was vastly more extreme than that of Southern California, arrived at the similar form by quite a different route. It seems clear from what he wrote at the time that the problem of heating during the hard winter occupied his mind more consciously than that of cooling the house in the hot humid summer that afflicts Chicago as much as other parts of the middle West. Nevertheless, stray references in writing, and observation of the houses themselves, show that he had evolved a method of designing coolness into the Prairie houses that went far beyond, say, the exploitation of existing vernacular methods or the imitation of any historical models.

But he seems to have started from an acknowledgement that

most of the houses his clients could afford would be built in normal lightweight construction, and that both winter and summer performance would depend on how he dealt with this fact. Yet it must be admitted that one can only say that he 'seems' to have started with such an acknowledgement of lightweight construction, because he has little to say about it in the Prairie house period. But this may be due to the fact that there was no compelling need to say anything at all because around Chicago everybody would take it for granted as a fact of life—to such an extent that he would not realise that it needed explanation to, say, a European audience. The other aspects of the construction and management of the Prairie houses he did perceive to need explanation, and the essential clues to his method of environment management are found in the text that he wrote to accompany the first European publication of the Prairie houses by Wasmuth in 1910. Thus:

Another modern opportunity is afforded by our effective system of hot-water heating. By this means the forms of buildings may be more completely articulated, with light and air on several sides. By keeping the ceilings low the walls may be opened with a series of windows to the outer air, the flowers and trees, the prospects, and one may live as comfortably as formerly, less shut in . . . it is also possible to spread the buildings, which once in our climate of extremes were a compact box cut into compartments, into a more organic expression, making a house in a garden or the country the delightful thing in relation to either or both, that imagination would have it.[8]

[8] from the English version of the explanatory text to the first Wasmuth volume (*Ausgeführte Bauten und Entwürfe*, Berlin, 1910) reprinted in Gutheim, *Frank Lloyd Wright on Architecture*, New York, 1914, pp 72ff.

Few writings of any architect relate mechanical equipment and plan and section so directly to aesthetic pleasure as does this compact and holistic vision of Wright's. Few statements of method can be so directly and revealingly tested against actual buildings. Although the statement begins with hot-water heating, it proceeds directly to the improvement of aspect and ventilation made possible by articulating the house into more separate parts, and in the process it inevitably involves lightweight construction on two counts. Firstly, a more massive type of construction, with better thermal insulation and higher thermal capacity, would make the actual

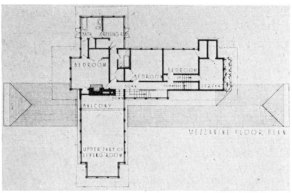

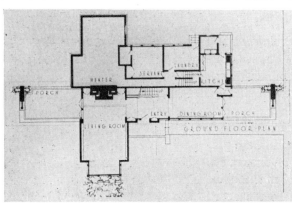

The Baker House, Wilmette, Ill., 1908, by Frank Lloyd Wright; above: street elevation; left; plans of main and upper floors; below: section throug bay window of living room.

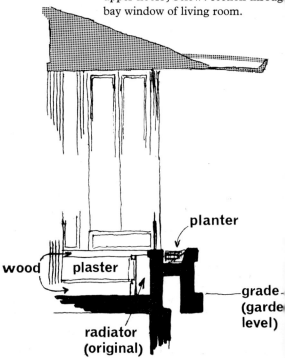

planter

wood plaster

grade (garde level)

radiator (original)

method of heating less crucial an issue, even if a highly articulated layout increased the surface through which heat might be lost. Secondly 'series of windows' implies large areas of glass, than which there is no more lightweight construction in common use in Western architecture. Even if the glass is carried between brick piers, as it often is in Wright's more expensive houses (Martin House, Robie House) its sheer extent gives the walling, on average, about as low a thermal performance as any domestic architect had dared offer up to that time.

However, it is in classic forms of wooden lightweight structure that the architectural meaning of Wright's words can best be examined, and the example to be cited here is chosen because it answers to these constructional norms, and has been studied by the author under extreme conditions of both winter and summer.

The Baker house, in Wilmette, Illinois, was one of the last Prairie houses to be completed before Wright, having fled Chicago in disgrace, wrote, in exile in Fiesole, the passage cited above—the relationship between words and architecture could hardly be more direct. The structure of the house is wooden studding and plaster. It looks (and is) flimsy, and has suffered some structural deformations as a consequence. The plan is not articulated to the extent of being broken up into isolated pavilions, but has the cruciform spread that is commonly identified as the hallmark of Prairie house planning. And it still contrives to temper 'our climate of extremes' satisfactorily, winter and summer, without the addition of any item or type of equipment that was not installed at the time of the original construction. The environmental performance is not 'good for its time', it is as good as the expectations of sixty years later would have it.

The manner in which it achieves this can be demonstrated by isolating the living room for separate consideration—while admitting that the whole house must ultimately be judged as an environmental unity. The living room is a large volume, thirty-two by seventeen feet on plan, and some fourteen feet to the highest point of the ceiling. The shallow pitched roof overhangs the side walls slightly and cantilevers a good way out at the south end, where it shelters an enormous bay window, some five feet deep by twelve feet wide. The glazing of this window is continued, as a clerestory strip below the eaves, some two feet deep and running back along both sides of the room, and thus encompassing the whole of its exposed perimeter.

The effects of this opening-up are subtle: the bay window is big enough to be used as a species of mediaeval 'solar', being provided with a permanent window seat on which one could sit to read or sew, and it looks out on what the immediate prospect has to offer in the way of flowers (growing in an external planter which, like the window-seat, is part of the permanent structure of the bay), trees and view. The room seems to be flooded with light from this vast window, but what in fact floods it in depth with light is the (often unnoticed) clerestorey strip, which brings light into the room on all sides. If the lighting of the room from the bay is thus somewhat of an illusion, its heating from the fireplace at the opposite end of the room is even more illusory. The hearth is little more than a visual effect, a sentimental symbol of home—the room is adequately heated, even in the depth of winter, by a hot water system of pipes running around the perimeter of the room concealed in the wainscoting, to serve a massive radiator under the window-seat, which is slatted to permit the warmed air to circulate.

The room is thus heated, and daylit, all round its three exposed faces, and the heating is neatly proportional to the probable heat loss through the glass—the biggest heating element is in the bay where the most glass is, the clerestorey being matched by the much smaller flow of heat from the feed-pipe runs in the wainscot. Such continuous heating is clearly a necessary provision in so lightly built a structure with a triple exposure to a bitter winter climate. But even in a cheap fuel economy (such as the US normally is) the resulting heat loss would be difficult to justify without some countervailing benefit. On this point, it should be remembered that

Wright had said that the houses were articulated not only for light on several sides but for 'light and air' and this is a point of the utmost importance.

The disposition of the windows in the Baker house living room is essential to the provision of adequate ventilation in summer. The bay window can be opened up almost entirely, as can two small windows in the clerestorey at the extreme north end of the room. Indeed, the seemingly gratuitous minstrel gallery over the fireplace at the north end of the room finds here a physical function, since it is the only means of access to these two lights to open and close them. Given their openability, and that of the bay, the breeze may blow across the room from side to side, or from end to end. More critically, from the point of view of the house as a whole, the two lights reached from the balcony are directly under the ceiling of the highest point in the house, and their opening allows hot air to escape exactly at the point where it is most likely to collect, while a cross draught through this point sweeps out the air that has been heated by contact with the warm underside of an exiguous roof-structure subjected to a Mediterranean sun (the latitude of Chicago is comparable to that of Naples).

The lessons of such a house are two, and interlinked. Firstly, that designing for mechanical services is not merely a matter of finding neat ways to install them—which Wright certainly did, in matters like the window-seat—but of setting them to work in partnership with the structure so that the whole is more than the sum of the parts. In the living room of the Baker house, plan and section, artificial heat and natural light, solid, void and overhang work together to give an equable indoor climate year in and year out, but work together in such a way that hardly any single detail participates in only one function, nor is any single function served by only one item of equipment—consider, for instance, the complex symbiosis whereby the parts of the bay window provide ventilation, shade, warmth, light, seating, greenery, view, flowers and privacy. But the second, linked lesson, is that this rich and

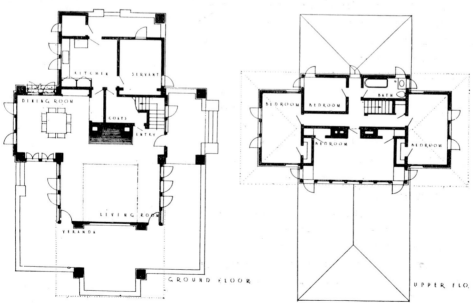

KITCHEN SERVANT

DINING ROOM

COATS

ENTRY

LIVING ROOM

VERANDA

GROUND FLOOR

BEDROOM BEDROOM BATH

BEDROOM BEDROOM

UPPER FLO.

improved environmental performance was achieved without recourse to any technological novelties—though Wright refers to hot water heating as a modern opportunity, it is the manner of seizing the opportunity that is modern, not the manner of heating, which was already a century or more old, and the manner of seizing the opportunity clearly has a great deal to do with the working together of the parts of the building. Here, almost for the first time, was an architecture in which environmental technology was not called in as a desperate remedy, nor had it dictated the forms of the structure, but was finally and naturally subsumed into the normal working methods of the architect, and contributed to his freedom of design.

A study even of the plans available in the standard literature will show how many of Wright's houses, from the Husser house of 1899 onwards, are developed from this approach to the management of the domestic environment—even though the standard literature is lamentably short on detailed environmental information, and the plans rarely show which windows can be opened and which not. Most strikingly, it emerges that the method is not essentially tied to very extensive or deeply articulated plans—even a compact little house like the Charles Ross cottage of 1902 shows Wright employing the same basic design techniques. This particular case is of interest because there is a work by one of Wright's major European contemporaries with which it can be compared. Peter Behrens's house for his own occupation in Darmstadt was built in the same year, has about the same usable floor-area, has a comparable heating technology (hot water pipes under window-seats) and its architect was almost exactly the same age as Wright and about the same fame in the land.

The Ross house is laid out on a rigorously axial modular grid, but contrives a remarkably free and open arrangement of the ground floor living spaces, while on the floor above, the three main bedrooms are quasi-cantilevered to give each of them a triple exposure and cross ventilation—as has the living room, though less obviously.

Facing page: the Charles Ross Cottage, Delavan Lake, Wis., 1902, by Frank Lloyd Wright; exterior and plans.

The Behrens house, lacking any over-all geometrical discipline, and being somewhat asymmetrical on every elevation, is nevertheless so completely constrained by pre-technological habits of planning and environmental management that the apparent freedom of its exterior proves to be no more than apparent. It remains a tight little traditional cubic box carved into smaller boxes. Some of the rooms do possess double exposures, but never with the generous window-provision—no less than 84% of the exposed wall of the Ross house living room—and panoramic views that characterise Wright's work. Compared to this, the Behrens house, in spite of its electric lighting and hot-water heating, is like a survivor from an earlier civilisation, the civilisation that had produced Catherine Beecher's Neo-Grec box of 1842, but not the American Woman's Home of 1869.

Wright clearly enjoyed extrinsic advantages that were not available to his European contemporaries—the stimulus of great environmental needs, the adaptability of lightweight construction, the resourcefulness of his contractors, the intelligence of his patrons and the remarkable cultural climate in which they lived in Chicago at that time, enriched by the philosophers and critics of the Chicago school around the University of Chicago, but not burdened by too ceremonial and European a concept of culture. But, even so, his greatest advantage was the intrinsic one of being some kind of genius and one of the most fluently inventive architects that ever lived. His resourcefulness in the deployment of power technology and structure together in the elaboration of domestic environments seems inexhaustible in the first decade of the century.

Thus, in the same years that the Baker house was being designed and built, he was also working on houses for Isobel Roberts, Mrs T. H. Gale and Frederick C. Robie. The Roberts house is a kind of toy miniature of the Baker house, complete with a gallery giving access to the only two clerestorey windows which can be opened, but introduces an innovation in the form of a version of the venti-

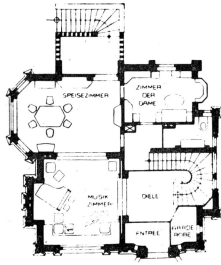

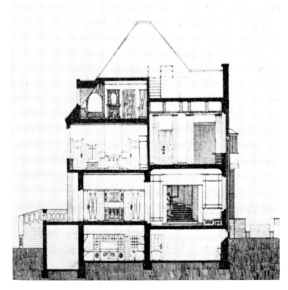

Architect's own house, Darmstadt, 1902, by Peter Behrens; top left: exterior; left: section; above: plans of main and bedroom floors.

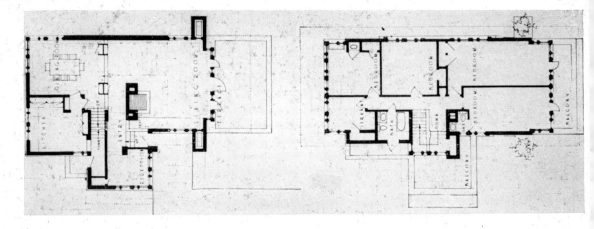

lating brick pattern device, here used to facilitate the flow of warm air from around the main hearth, as if he was busy re-inventing the hot air stove from first principles. The Mrs Gale house, another studding and plaster construction with very generous fenestration, represents another approach entirely. To outward view, it appears to consist of superimposed floor-slabs, supported by massive buttresses on either side, but from inside the living room one sees that these seeming buttresses are hollow, and open toward the living space. Within these curious cupboards are hot-water radiators placed edge-on to the room and concealed by slatted wooden grilles.

There is clearly something wrong with the manner in which these radiators are installed in these 'servant spaces', especially as they have had to be supplemented by a large free-standing radiator alongside the fireplace. The floor-space occupied by this radiator was reserved on the published plans for a built-in bench, but there is no sign that such a bench was ever made or fitted. The solution to this mystery appears to be that the house was originally conceived in terms of hot air heating, for which the pseudo-buttresses were to house the riser ducts for the living room and bedrooms above, as were the mysterious free-standing cupboards

which now occupy part of the steps leading up from the living area to the dining space.

Gale House: exterior from the street.

This, admittedly, is conjectural; so are some of the innovations of the Robie house, though others are beyond dispute. This last, and finest, of the Prairie houses has been so much discussed and published that there would seem to be very little left that need be said about it, but its environmental innovations have received hardly a footnote in the standard literature. Yet the authors of the standard literature can hardly have entered the house without being struck by environmental peculiarities (to rank them no higher) worth mentioning—all we have is Giedion's mysterious

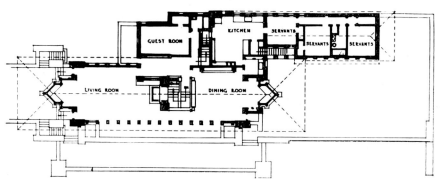

The Frederick C.
Robie House,
Woodlawn
Avenue, Chicago,
Ill., 1910, by
Frank Lloyd
Wright; above:
the south front;
left: floor plans.

GUEST ROOM · KITCHEN · SERVANTS · SERVANTS · SERVANTS

LIVING ROOM · DINING ROOM

UPPER FLOOR

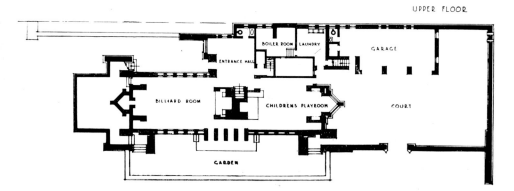

BOILER ROOM · LAUNDRY · GARAGE

ENTRANCE HALL

BILLIARD ROOM · CHILDRENS PLAYROOM · COURT

GARDEN

complaint that like other Prairie houses it is dark, because of the overhanging eaves.

Yet the first environmental feature of the house strikes the summer visitor even before he is through the door—the coolness of the shaded entrance court. It lies on the north side of the house and receives practically no sun; together with the basement floor, also profoundly shaded by the projection of the terrace above its windows on the sun-ward side, it provides a cool-air tank that works so efficiently, even on still thundery days of high humidity, that with all the windows shut even the highest rooms under the roof are not uncomfortably warm.

This cool-air tank effect seems to be unique in Wright's work, and the topic is not one that he seems to have discussed, so it may be that this is an accidental environmental benefit. The other benefits however follow the general lines of the methods already discussed, while bringing in some fresh ideas in connection with the use and exploitation of electric lighting. Because of the large areas of glass—the main living floor is exposed, and fenestrated, for more than three-quarters of its perimeter—this is effectively a light-weight structure, in spite of being built of brick piers with 15 inch welded steel joists in the main roof to support the sweeping overhangs to east and west. Though divided in the centre by a stairwell and a ceremonial fireplace, the main floor is effectively a single space, differentiated functionally into a living end and a dining end. Practically every square inch of fenestration, protected by internal flyscreens, can be opened, so the profusion and variety of ventilating conditions is generous indeed.

Nearly every single openable light is matched by a heating element: under the half-depth windows overlooking the entrance court to the north, radiators standing on the floor-level are neatly boxed in behind slatted gates; there are hot pipes at the backs of the built-in cupboards in the bay windows at the ends of the room, with slots in the skirting and the cupboard-tops to permit the warmed air to circulate; and provision was made for another set

The Robie House: left: part of the living room, showing lights and ceiling grilles; opposite, section and part-plan of living floor showing environmental provisions.

1. Roof overhangs to s and w
2. Opening windows
3. Glazed doors
4. Roof space
5. Radiators under windows
6. Radiators sunk in floor
7. Glass lighting globes
8. Structural steel beam
9. Dimmer-controlled bulbs
10. Lighting grilles
11. Hinged fly-screens

of dwarf radiators, matching those under the windows to the north, to be sunk in the floor under a brass grille at the threshold of each of the french-windows that open out onto the terrace to the south.

Whether these radiators were ever installed is not now clear. There is room enough for them, and there is a massive hot water pipe in the bottom of the cavity provided for them, but they are not there now. It seems that the house may not have been one of Wright's successes thermally, and it has a reputation for being hard to heat. But this, of course, might have been due to a failure to install the radiators originally envisaged. At all events, later occupiers than the Robie family have installed free-standing radiators in the corners of the main floor, in locations that patently are not of Wright's choosing.

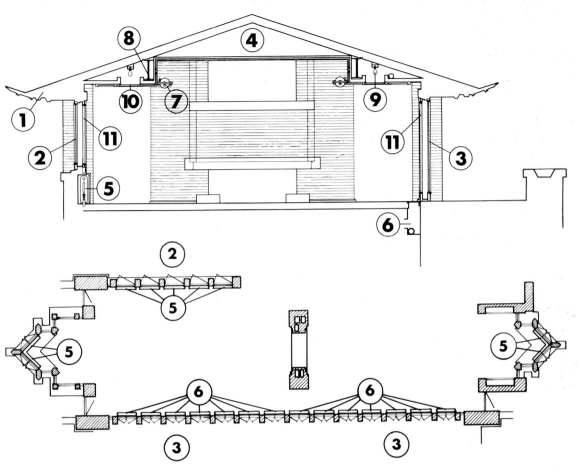

The lighting system, as it survives, is entirely his, however, except where decorative details have been knocked off in the passage of time and use. And it is one of the triumphs of the early history of the art of electric lighting. While local light was provided where it was needed by special lamps of some elaboration built in to Wright's purpose-designed furniture (eg., on top of the corner piers of the dining table) the more general and overall lighting of the main living floor was provided by an installation that was an integral part of the fabric and conception of the building. The immediately visible part of that lighting was provided by the frosted globes carried in square japonaiserie frames which occur on the edge of the central upstand of the ceiling, at points corresponding to the structural piers of the outer wall. But corresponding to the glazed openings of the outer wall, and inset in the lower flat portion of the ceiling, are a series of oaken grilles, made into abstract designs by the insertion of small cubes of oak between the slats of the grille. An electric lamp, controlled by a dimmer, is set above each of these grilles, and casts a modulated dappled light through it onto the floor in front of the windows.

However romantic the effect may be judged, there can be no doubt that this is an entirely appropriate installation, in a situation where the customary sconces or wall brackets would have made little visual sense. But these oaken grilles may have a further function which is nothing to do with lighting, to which the clue is given, once more, by the text of the Wasmuth portfolio:

> The gently-sloping roofs, grateful to the Prairie, do not leave large air-spaces above the rooms, and so the chimney has grown in dimensions and importance, and in hot weather ventilates at the high part the circulating air-space beneath the roofs, the fresh air entering beneath the eaves through openings easily closed in winter.[9]

[9] Wright (Gutheim), loc. cit.

The Robie house has such openings beneath the eaves and its chimney has grown a separate limb projecting on its western side, with the characteristic pattern of ventilator bricks indicating an extract of some sort. But if, in winter when the openings under the eaves had been 'easily closed,' air was still required to circulate through the roof-spaces to carry away the damp, it could enter through the lighting grille, acquiring useful heat from the lamps and from sundry hot pipes in the roof. There would then be just room for it to pass between the flange of the steel joists and the underside of the roof covering, and so pass into the higher part of the living-room roof above the raised ceiling, and thence out by way of the ducts in the brickwork of the added limb on the side of the chimney.

The great cantilevers of the roof, while they contain no hidden environmental secrets, are not quite so simple in their function as might appear. The western one certainly functions as a sunshade, particularly against the raking mid-afternoon sun which Latin cultures call 'the light that kills,' but the eastern one probably does more useful work as a permanent umbrella to keep the rain off the kitchen and service entrance area in the corner of the car-yard. Some have seemed surprised that the projections of the eaves on the south side of the house seem slight compared with those to east and west, and doubt that it is sufficient to shade the glass. In fact it is *exactly* sufficient—the sun stands tall in summer in Chicago's latitude, and at mid-day on Midsummer day, the shadow of the eaves just kisses the woodwork at the bottom of the glass in the doors to the terrace.

Whether this nicety is of any real value or was even deliberate is now impossible to tell—it would be a fitting flourish (too fitting for real life) on which to end Wright's greatest period of architectural and environmental mastery. Nothing he did in the rest of his long career quite matched the inventiveness and total control over every aspect of the building that characterises the Prairie houses of 1899–1910. Nothing else that anybody else did for decades was to match their easy mastery of environmental control. With Wright's departure from Chicago, the subsequent splintering of the Prairie school, and the contemporaneous running down of the California school (though this was to find surprising successors, as will be shown in chapter 10) the architecture of the well-tempered home passed its first peak; the second was to be some way off, and the intervening period was to be largely occupied, and profoundly confused, by the European modernists' conviction that 'the Machine' was a portentous cultural problem, rather than just something that the architect could use to make houses 'perfectly sanitary, labor-saving . . . where the maximum of comfort may be had with the minimum of drudgery.'

7. Environment of the machine aesthetic

Whereas Wright took over relevant technologies with enthusiasm but without dogma, applying them as required but without prejudice to stylistic preferences, the European modernists who passed, or claimed to have passed, under his influence could not handle the matter in so cool and relaxed a manner. Years of indoctrination since Ruskin had made technology a problem, not an opportunity, while rationalist theory such as was made explicit in the *Histoire* of Auguste Choisy, relating changes in style to changes in technique, led them to look for some new style as an automatic product of the new technology. Their pre-occupation was to fix the correct new style of the Machine Age, and thereby resolve the problems that it threw up. As Marcel Breuer put it in 1934:

> The origin of the Modern Movement was not technological for technology had been developed a long time before ... What the new architecture did was to civilise technology.[1]

Such a statement is not one that Wright could have made in the Prairie house period, nor could any European modernist during the confident twenties, nor in the later thirties, when the tone of discussion was to become moralising, and deterministically Functionalist, maintaining that the public must accept modern architecture (*scil.* International Style) because it was historically necessary in a technological culture. But in the difficult pause of 1934, at the beginning of political exile, Breuer could look back and speak with commendable frankness of the struggle to persuade the public that they ought to accept the asperities of the Machine Age style along with the comforts and conveniences of technology:

> ... it proved harder to formulate the intellectual basis and the aesthetic

[1] Breuer, lecture given in Zürich, 1934; English text in *Marcel Breuer, Buildings and Projects*, London, 1962.

of the New Architecture, intelligibly, than to establish its logic in practical use. One has experienced all too often that something like a functional kitchen equipment has made hypercritical people more accessible to our ideas, and that as a result they have not infrequently become reconciled to our aesthetic. [2]

[2] ibid.

One cannot help suspecting that even less-than-hypercritical people were puzzled to know why they were supposed to endure one set of environmental discomforts in order to enjoy another set of comforts when no-one had formulated 'intelligibly' any reason for doing so. Breuer is distinctly defensive about the discomforts that the cultural preferences of the architects (those ideal and universal abstractions of which Oud had spoken) had enjoined, and the phrasing of his argument is curious, if familiar.

> Are our buildings identifiable with descriptions such as 'cold', 'hard', 'empty looking', 'ultra-logical', 'unimaginative and mechanistic in every detail'. . . Whoever thinks so has only seen the worst examples of modern architecture, or has had no opportunity to live in or make a close inspection of the best. Does he perhaps mean 'pompous' when he says 'human', a brown sauce of wallpaper when he invokes cosiness, empty pretence when he demands peacefulness ? [3]

[3] ibid.

The way in which questions which are susceptible of straight-forward physical investigation are nudged up to the 'higher' plane of cultural problems, so that they may be dealt with as manifestations of mere old-fashioned prejudice, is extremely striking. Thermal coldness and acoustic hardness are both susceptible of at least comparative measurement; the tempered reflections of light from brown wall-paper may indeed be more objectively 'cosy' than the ophthalmic irritation of glare from glossy white surfaces. It is difficult to avoid the impression that Breuer is now being less than frank, and attempting to double-talk his way around some solid environmental objections.

For it is now clear that there were some real environmental deficiencies to object about in the 'White architecture of the Twenties'. More than one loyal supporter of the Bauhaus at that time is now, forty years later, prepared to admit that the glaring lighting and the ringing acoustics were distressing, and Philip Johnson has told the author that he had difficulty in remaining in the Bauhaus buildings at Dessau for more than about forty minutes. Where other interiors of the period survive, or have been restored or reconstructed, one may still encounter these trying conditions, but in most cases the addition of carpeting, curtains, and light-fittings not visible in the early photographs have rendered these interiors easier to inhabit.

It was in Germany, almost more than anywhere else, that the promise of improved environmental quality was most ruthlessly sacrificed on the altar of a geometrical machine aesthetic and the honest expression of everything, including the sources of light. It was an extraordinary retreat from the humane concepts of environmental quality that had been evolved before 1914. As Michael Brawne expressed the matter thirty years after Breuer had spoken, the blame lay upon

> . . . the Bauhaus, which completely neglected the consumer, and concentrated on how most rationally to produce the object—how, in the simplest way, to blow a globe of glass which then becomes a light fitting.[4]

[4] reported in *Architectural Association Journal*, March 1960, p 155.

without regard for the glare which it throws into the eyes of the occupants of the room. Although a similar disregard for the physiological responses of the consumer can be seen among the modernists in France in the same years of the twenties, and will be discussed in the next chapter, the retreat from comfort, from pleasure and from creative imagination is the more stunning in Germany because there, uniquely, a great testament to the art and pleasures of lighting had been published just before the outbreak of the First World War.

In many ways this document is not unique in its intellectual roots. A Futurist-inspired belief in a better environment through the exploitation of machine technology had been fairly widespread in the nineteen-teens. The boss Futurist himself, F. T. Marinetti, refers to it more than once, demanding at one point

... well ventilated apartment houses, railways of absolute reliability, tunnels, iron bridges, vast high-speed liners, hillside villas open to the breeze and the view, immense meeting halls and bathrooms designed for the rapid daily care of the body.[5]

and at another invoking the fortunate engineers who will enjoy

... a life of power between walls of iron and crystal; they have furniture of steel, twenty times lighter and cheaper than ours ... Heat, humidity and ventilation regulated by a brief pass of the hand, they feel the fullness and solidity of their own will.[6]

[5] lecture to the Lyceum Club, London, 1912; reprinted in *Le Futurisme*, Paris, 1912.

[6] ibid.

When this was written in 1912 it was still pretty speculative stuff, although most of the necessary technology existed somewhere by that date. But the very artificiality and abstract language of the vision of the future environment deprive it of the quality of experience which gave earlier Futurist writing such impact—an impact which survives in the writings of Paul Scheerbart, and especially his book *Glasarchitektur* of 1914.[7] Of all the visionary writings of that period, this book has the greatest impact nowadays as the concrete and tangible vision of the future environment of man.

[7] Paul Scheerbart, *Glasarchitektur*, Berlin, 1914.

Scheerbart has a small assured place in the history of German literature for his witty and fantasticated novels of scientific and erotic fantasy. In nearly all of them there is an extraordinary power to make the reader see and feel what is being described, and this power seems to derive from a remarkable capacity for direct observation, enriched by a great variety of experience occasioned by Scheerbart's knockabout existence as a Berlin Bohemian. One of the key experiences he observed was the horror of being poor and cold in a world of brick and masonry;

Backsteinkultur tut uns nur Leid

he wrote, and went on to berate the *Backsteinbazillus* which he declares to inhabit old masonry. While this private demon is clearly even more imaginary than the 'organic poison' which the Victorians had supposed to vitiate air, an intelligent and perceptive slum-dweller of the period would not need much imagination to invent it where the characteristic and depressing odours of dry rot and damp plaster hung on the poisoned air.

Against the horrors of *Backsteinkultur*, Scheerbart set his vision of a future architecture of glass and steel and light. Unlike so many visionaries, however, he had an acute consciousness of the mechanical and environmental realities of the architecture he was proposing. He does, admittedly, begin by placing his vision in a cultural context:

> We mostly inhabit closed spaces. These form the milieu from which our culture develops. Our culture is an exact product of our architecture. If we wish to raise our culture to a higher plane, so must we willy-nilly change our architecture. And this will be possible only when we remove the sense of enclosure from the spaces where we live. And this we will only achieve by introducing Glass Architecture which will let the sunlight and the light of the moon and stars shine into the room, not through a couple of windows but, as nearly as possible through whole walls, of coloured glass. The new milieu so created will bring us a new culture.[8]

[8] ibid., p 11.

As was said above, such visionary beliefs were not exactly rare at the time, but few of the others were based on detailed observation of new environments in actual use. Early in the book, Scheerbart comments on the glass-houses in the Botanical Gardens at Dahlem, admiring the spectacular effects of the evening sun as seen from inside them, but regrets that they are not double-glazed to prevent excessive heat-loss. And here he produces—as few other authors would have done then—the figures: on a day that starts with an 8 a.m. temperature of $-10°$C, the heating plant consumes *300 Zentnern bester Schlesische Steinkohle*. Furthermore, he knows how and why double-glazing works, and how it could be exploited aesthetically:

> Since air is one of the worst conductors of heat, all glass architecture needs this double wall. The two glass skins can be a metre, or even farther, apart. Lights between these walls shine both inwards and outwards, and the outer, as well as the inner glass can have coloured ornamentation. If too much light is then lost in the colours, the outer skin can be left clear, and all that is needed is a coloured glass screen

between the two skins to temper the light, so that the outward light in the evening does not appear too plain.

Convectors and radiant heaters should not be put between the two skins because too much of their output will be lost to the outside air.

Obviously glass buildings are only suitable for building around the temperate zones, not in polar or equatorial areas. In the torrid zones we will not succeed without a sheltering white concrete roof, but there is no need for such roofs in the temperate zones.

And for warmth, one could grace the floor with electrically heated carpets.[9]

[9] ibid., p 14.

These observations not only show informed common-sense, but also a keen understanding of where technology could be made to go next—the first electrically heated textiles did already exist (General Electric in the US had catalogued a Warming Pad in 1906) but it is doubtful if such techniques had been seriously applied to floor-warming at this date. Nevertheless it is upon the double basis of shrewd anticipation and available common experience that Scheerbart sets out to persuade his readers that the change to the new 'milieu' which is to be the environment of the new culture, will be no problem: *Der erste Schritt ist sehr leicht und bequem!* He starts from the glass architecture that most of his readership would know from domestic experience, the glass-roofed veranda, and points out that it

> . . . would be easy to enlarge, and then surround on its three sides with double-glazed walls . . . and if you wish for a view of the garden, that can easily be had through inserts of window-glass. But it would be well not to arrange these in ordinary window form. Air is better brought in through ventilators.[10]

[10] ibid., p 12.

and later he suggests that the veranda, thus enlarged, should be detached from the house entirely, and set up as an independent structure in the garden.

His view of the glass future is very complete and detailed. There will have to be new materials to replace wood for glazing bars, and new surface treatments and claddings for concrete, which he finds an unsatisfactory material as it comes off the form. The colours and ornamentation of glass architecture can be derived from mineral and vegetable forms, or from the technology of glass itself—he mentions the smearings and marblings of *das Tiffany-Effekte*. Furniture of steel, upholstered in cloths of glass-fibre (a technique 'now forty years old', he claims) will be grouped in the centre of the rooms, so as not to mask the beauties of the glass walls. Light is to stream from illuminated columns or from under glass floors, or from floodlights playing on the outside of the building. Alternatively, the building will be seen externally as light modelled in forms—*ein ganz selbstandig Illuminationskörper*, a complete and independent body of illumination.

And so this extraordinary vision continues, throwing off neat innovations like using the vacuum cleaner to remove garden pests, foreseeing the disappearance of the word *Fenster* from the dictionary as the wall becomes all glass; visualising a landscape that anticipates *Son et Lumière*—floodlit mountains, glass buildings reflected in lakes, the night turned into day and into a great work of art. Such prophecies always suffer the normal entropies of passing time, and much of what Scheerbart prophesied has come true in oblique ways he could not have anticipated; in the nightscape of Las Vegas, for instance. But much was to come directly true, and none of it more tellingly true than his anticipations of immediate error to be made by architects in their use of light. Indeed, it is the bedazzled inhumanity of the lighting of the 1920's that now makes the sanity of his views so clear.

It will have been apparent from what has been quoted already that Scheerbart was set in almost exactly the opposite direction to that modern architecture was about to take, even if its practioners shared his passion for glass. Not only does he pooh-pooh the use of their favourite material, white concrete, in the climates where they normally used it, the temperate zones, but at another point in the book he specifically attacks 'the so-called *Sachstil*', by which he clearly means the stripped-down architecture of the Deutscher Werkbund. This, he says, is acceptable only in a transitional period as a means of eliminating the traditional styles of *Backsteinarchi-*

tektur und der Holzmobel, and that afterwards the architecture of glass would evolve its own rich *Ornamentik*—or, at least, *Hoffen wir's!*

This hope was not to be fulfilled to any significant degree, but the implied criticism of the lighting techniques of the *Sachstil* that emerge in the book were to prove entirely justified. The stand that Scheerbart adopts looks startling at a time when almost everybody, and not just the electrical industry, seemed bent upon greater and greater levels of illumination, but later events, and the research methods developed for scientific lighting studies seem to justify almost every word of arguments like these:

> When we have light in plenty, as the fuller exploitation of wind and waterpower will make possible, so we will have less need to leave the light clear, and will be able to temper it with colours . . .[11]

> Full white light is the cause of at least part of the neurosis of our time. Colour-tempered light settles the nerves, and is used by neurologists as an element in the cures at their sanatoria.[12]

> Thus, an increase in the intensity of light is not what we need. It is already much too strong, and can no longer be tolerated by our eyes. Tempered light is what we need. Not 'More light!' but 'More coloured light!' must be the call.[13]

[11] ibid., p 63.

[12] ibid., p 103.

[13] ibid , p 120.

In the end, pure white light was to survive only as the weapon of the Secret Police interrogator, the brain-washer and the terrorist. But before that relegation to the underworld of Western culture, it had almost a two-decade career in the visible and progressive overworld, as architects of the International Style—with the noblest aspirations, and clear consciences which the clarity of the light was supposed to symbolise, no doubt—subjected doctors, artcollectors, publishers, teachers and the other law-abiding bourgeois who were their clients, to a Gestapo-style luminous environment, with light streaming from bare, or occasionally opalescent, bulbs and tubes and glaring back from white walls. Even when allowance is made for the fact that many of the interiors they designed were for the specialised purposes of exhibitions, and may have needed unusual intensities of lighting, the published record of the work done by the Bauhaus and like-thinkers down to 1934, combined with the memories of survivors, leaves an impression of a luminous environment close to the threshold of pain, probably made tolerable only by the notorious willingness of intellectuals to suffer in the cause of art.

Just before this bleak interlude began, however, there was a brief period when Scheerbart's common humanity and uncommon romanticism about the use of light and glass, looked as if it might take hold and inspire some architecture to match. The direct influence of *Glasarchitektur* on the work of the Berlin Fantasists (most of which was doomed to remain on paper, anyhow) can be seen with certainty only in the work of Bruno Taut, but the three years after the Armistice did produce a spate of projects for irregular glass towers—even from so grave a rationalist as Mies van der Rohe, as witness his submission in the Freiderichstrasse competition, and its derivatives. Again, the interiors sketched by Hermann Finsterlin and those actually built by Hans Poelzig in the Grosse Schauspielhaus, also stand close to Scheerbart's ideas.

But in the case of Bruno Taut, the influence is both visible and acknowledged. *Glasarchitektur* was dedicated to him, and Taut's glass pavilion at the Werkbund exhibition in Cologne was dedicated to Scheerbart, who performed the opening ceremony. The pavilion was glazed in roof and wall, floor and stair-risers, was devoid of rectangles in plan or section, and much of the glass was coloured. It was palpable proof that Scheerbart's vision could be realised, even in the year in which the book was published, though the design must have been in hand at least while the book was being written, because Scheerbart makes references to Taut's difficulties in finding a suitable material in which to model it.

Even if their acquaintance had begun in 1913, the association of Taut and Scheerbart must have been short-lived, because he died in 1915, and from then on his direct influence begins to wane, even in the work of Taut. Nevertheless, Taut treasured the memory deep into the twenties. Not only are there strongly Scheerbartian

Glass Industry pavilion, Cologne exhibition, 1914, by Bruno Taut; left: the main exhibit room; top right: the exterior, and right, one of the glass staircases.

visions in the pages of his magazine *Frühlicht*, but a whole short story of Scheerbart's is quoted as an introduction to *Die Stadtkrone*, which appeared in 1919.[14] More striking than this, however, is the fact that the influence can still be detected when Taut published a book on his own house in 1927. In spite of its eccentric form, the house bears little direct resemblance to the glass pavilion (though it does employ glass bricks in the staircase wall) but at one or two points, both in what was done and in how it was described in *Ein Wohnhaus*, Scheerbart's accents can still be discerned, in however attenuated a form. Thus the special lamp over the dining table, according to Taut

[14] Jena, 1919.

> . . . illuminates strongly the table-top, which does not dazzle because it has no white cloth on it. The light itself does not dazzle either, but glows gently from the lamp (of Luxferprisms) and is hardly reduced in intensity at all, as it would be by opal glass or cloth drapery. Against the black plane of the table gleam crockery, glass and flatware—but one can no longer call it a table setting, for the table has vanished![15]

This delight in glitter and light and the dematerialisation of traditional solids (the table surface) clearly stands in Scheerbart's tradition, and so too does the interest in demonstrative colour. At a time when German modernists were forgetting the colour lessons they had learned from van Doesburg, Mondriaan and other members of *de Stijl*, and were well on the way to their ultimate aridities of *Weiss-im-weiss*, nickel plate and black leather (epitomised in Gropius's designs for the Paris exhibition of 1930) Taut was using colour demonstratively throughout the house, and even publishing a sample colour-card in *Ein Wohnhaus*, so important did he take it to be in the design.

In one or two cases the colour-schemes of actual rooms are given (or can be deduced)—for instance, in a red room the heating pipes and radiator are picked out in yellow. This is what was meant, above, by using colour 'demonstratively'; more than anything else, Taut uses his colours to draw attention to his mechanical equipment. He was clearly proud of his gadgetry, down to the last

Facing page: Architect's own house, Berlin 1926, by Bruno Taut; top left: working corner of the master bedroom; top right: central roof ventilator; bottom left: dining table illuminated; bottom right: light fitting for dining table.

[15] *Ein Wohnhaus*, Stuttgart, 1927, p 51.

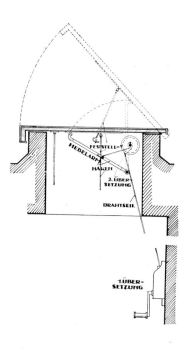

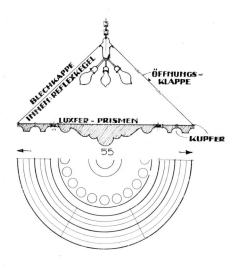

ventilator handle; it is all catalogued in the book, and rendered rhetorical in the house by the use of colour which makes it everywhere obtrusive in the room spaces. Colour apart, this puts Taut more in line with the aesthetics of the age, which in Europe dealt demonstratively with mechanical kit, even when its useful performance had barely been exploited through any reasonable adaptation of architectural form to seize the opportunities offered by such services.

Thus, some of the neater touches in Gropius's work of the late twenties consist of the apt deployment of things like heating equipment—apt, that is, within the narrow dogma of the ruling aesthetic. In the Employment Exchange for the city of Dessau, for instance, the hot-water radiators on the edge of the floor-slab at each landing of the main staircase serve not only to temper the chill of a multi-storey window forming the whole of one wall of the stair-well, but also double as balustrades to prevent persons walking through the glass.

Facing page: employment exchange, Dessau, Germany, 1927, by Walter Gropius; above: stair-tower and administrative offices; below; glazed ceiling and electric lights in main hall.

This humane and intelligent usage should not, however, be seen as inconsistent with the methods of lighting to which reference has been made above. For Gropius and the Bauhaus connection, lamps and heaters alike seem to have been simply sculptural objects, to be composed according to their aesthetic rules, along with the solids and voids of the structure, into abstract compositions. As for their environmental performance, this seems to have been honoured only by the observance of certain simplified rules (or, possibly, eroded habits of mind) by which the radiators were placed flat against the outside walls, and the lamps hung from the centre, or the centre-line, of the ceiling. Where there is any conspicuous departure from such a rule, it usually emerges as a purely aesthetic 'improvement' without regard to its environmental consequence, or a desperate attempt to remedy an environmental mistake already made.

The office which Gropius designed for himself in 1923 is a textbook example, partly because it is already in the text-books, and

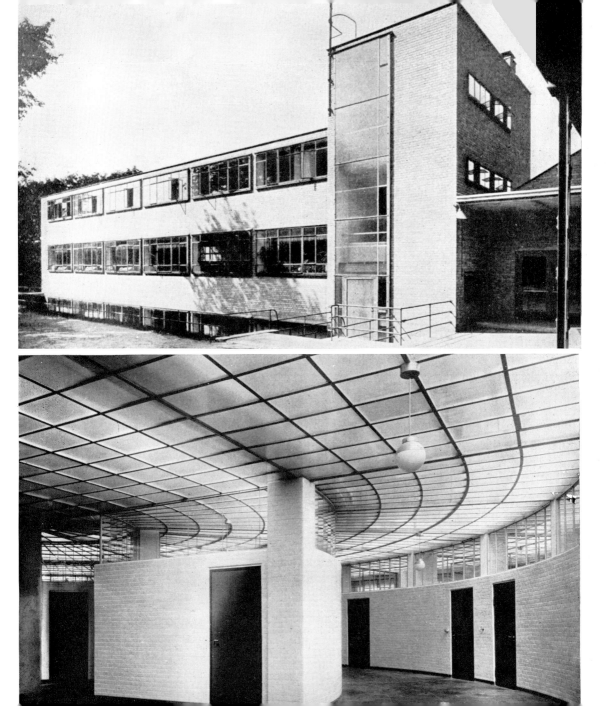

partly because it neatly exemplifies both tendencies. The room was to be conventionally heated, by a radiator which is clearly shown in the well-known isometric drawing; and it was to be unconventionally lit, by an abstract pattern of tubular lamps suspended unshaded from the ceiling as part of an Elementarist space-sculpture, two of the tubes being near the side-walls and related to a change in their colour and texture, and the other four in a complex cluster somewhere near the centre of the ceiling. The controlling discipline that relates all these elements is visibly an aesthetic one, concerned with converting an existing room into a modern space-composition, without much regard to the human use to be made of it. Proof of its environmental and human inadequacies is afforded by later photographs of the completed room, in which a table-lamp, not specified in the original design, has appeared on the desk to deal with the deficiencies of the lighting originally designed into the room, together with two telephones on a shelf (also missing from the original drawing) next to the radiator. It all suggests that the 'New Synthesis of Art and Technology' which Gropius was preaching at the time could contain oversights that left it less than the total discipline of design with which his followers have credited him.

Facing page: design for interior of own office, Weimar, 1923, by Walter Gropius.

This room-interior has been criticised before, of course, on the grounds that the lighting fixture was copied, and copied without comprehension, from that which G. T. Rietveld had designed for the study of Dr Hartog at Maarssen in 1920. While it is true that Rietveld's version, regarded purely as a piece of sculpture, represents a more advanced spatial composition, neither has much to recommend it as a source of light for a room inhabited by human beings. Nevertheless it is instructive to compare how these two symptomatic European designers, Gropius and Rietveld, developed their attitudes and practices and lighting through the twenties and early thirties.

Rietveld's contributions to the electrical and electronic environment include, for instance, a neat and original table lamp, which

Light fitting in the study of Dr Hartog, Maarssen, Holland, 1920, by G. T. Rietveld

136

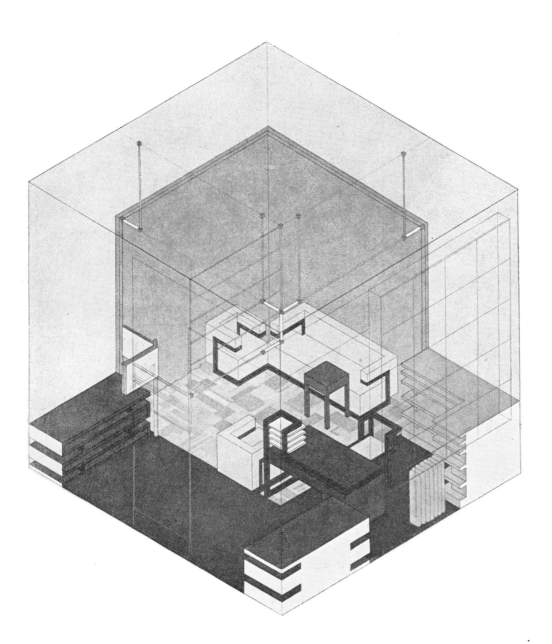

also appears as part of the extraordinary fully-glazed and mobile radio cabinet he designed for the Rademacher-Schorers in 1925. In neither case, however, does Rietveld appear to have operated as a man making an improvement in the acoustical or luminous environment, but more as a sculptor composing given mechanical forms into a work of art. Other interiors of the ensuing years show him accepting such inhuman conventions as the naked lamp-bulb, however subtle or striking might be his use of daylighting in interiors like those of the Schröder house in Utrecht. Where the contrast is even stronger, however, is between his artificial lighting of interiors and that of exteriors. In a series of shop fronts he used lights and illuminated lettering to create virtual volumes of luminous space that would be difficult to parallel in any European

Radio-cabinet and table lamp, both 1925, by G. T. Rietveld.

Bioscoop Vreeburg cinema, Utrecht, Holland, 1936, by G. T. Rietveld; the illuminated façade at night.

architect's work of the period, and crowned this achievement with the stunning façade of Bioscoop Vreeburg (Utrecht) of 1936, which is almost the only thing that Europe has to set against the sophistication of cinema façades in the New World by that time.

At the Bauhaus, the output of environmental hardware in the form of light-fittings etc., was prolific, but all are subject to Michael Brawne's objection, cited above, that they were designed for ease of production, not quality of illumination. Almost the only exception to this rule was a small bedside lamp, now of disputed authorship and date (originally given as 1924) with an adjustable shade, which could still pass muster today. Aesthetically, and mechanically, the design of light-fittings (especially those by Marianne Brandt 1925–1928) did become increasingly sophisticated, but there seems

Row-houses, Utrecht, Holland, 1930, by G. T. Rietveld; left: exterior showing modified window of corner room; right: corner room showing illuminated, lowered ceiling.

to have been little matching interest among the architects in the use or exploitation of the lighting-effects that might have been created with them. Though there are a number of projects for lighting-displays for purposes such as advertising in the published literature on the Bauhaus, the organisation did not, for instance, advertise itself by illuminating the large 'Bauhaus' logotype that ran down the end wall of the engineering block at Dessau.

Again, reverting to interior design, it is striking that in the Employment Exchange building at Dessau, an elaborate system of laylights and translucent ceiling panels is used to diffuse daylight into windowless interior spaces on the main floor, but the same system is not used to diffuse the artificial light, which still comes from the inevitable translucent globes suspended clear along the centre-line of the space below the translucent ceiling. It seems doubtful indeed if any of the European modernists who are generally accepted as being in the mainstream of descent from the true pioneers of modern design, ever fully realised the possibilities of concealed lighting in ceilings, in the way that their less reputable brethren, discussed in chapter 9, did—with one notable exception. That is Rietveld's use of a lowered, luminous ceiling in the corner house of a terrace in Utrecht built in 1930–1931. Even here, the strict aesthetic morality that would normally make it mandatory to exhibit such light-sources as bulbs or tubes, insists on a compensating honesty in the *external* appearance of the building, expressed as the suppression of the row of opening lights that form the upper parts of the other main windows along this façade, and their replacement by solid walling, to close the end of the space above the suspended ceiling.

But this use of an electrically lit luminous ceiling is an exception to a generally unimaginative use of artificial light which is made the more baffling by the fact that at the Bauhaus, for instance, certain non-architectural departments were making the most radical

Pantomime with Figures and Translucent Walls, Bauhaus Theatre, 1927.

141

explorations in the use of light. While the architecture and industrial design departments were refining the equipment for their
inquisitorial conception of interior lighting, the sculpture work-
shop, the designers of temporary settings for festive occasions and,
above all, the stage department, showed a command of trans-
parency, modelling, virtual volume and space-definition through
light effects, that would have delighted Scheerbart, and would have
had an immensely stimulating effect on any architectural department that was not so rigidly restrained by its own conception of
the aesthetic necessary to 'civilise' technology.

Light-play, Bauhaus Theatre,
1927.

8. Machines à habiter

Le Corbusier's position of unrivalled esteem among architects makes him too convenient a target for criticism, too obvious a colossus on whom to find feet of clay. The temptation to find him the most signally delinquent of his generation in matters of environmental management should be resisted. Even though the implicit and explicit promises of his writings lay him open to more damaging attack, he was probably no worse than the rest of his generation, the rest of the connection who, under the guise of the *Congrès Internationale d'Architecture Moderne*, became the official Establishment of architecture in our time. The whole generation was doubly a victim; firstly of an inability of its apologists and friendly critics to see architecture as any more than a cultural problem, riding upon a conventional view of function that had not been related to twentieth-century needs; and, secondly, of its own (apparently willing) submission to a body of theory more than half a century behind the capabilities of technology, still preoccupied with problems—such as the use of metal and glass in architecture—that had been propounded by the generation of Sir Joseph Paxton and Hector Horeau in the 1850's, and so effectively solved by those mid-Victorian masters that the practical results were common knowledge for those, like Paul Scheerbart, who cared to seek them out at first hand.

For those who had (or were content) to take such experience at second hand from books, the situation was less happy. The available texts, if one may judge by, say, Richard Flügge's *Das Wärme Wohnhaus*,[1] still dealt entirely with conventional constructions of load-bearing brick, pitched roofs of tile or slate, etc., and contained little information that could be applied to lightweight structures of concrete and glass. Nor was the available historical record any

[1] *Das Wärme Wohnhaus*, Hall, 1927. Its attitude seems the more unenterprising in view of Flügge's original contributions to environmental knowledge; see the quotation from Kimball in chapter 2.1.

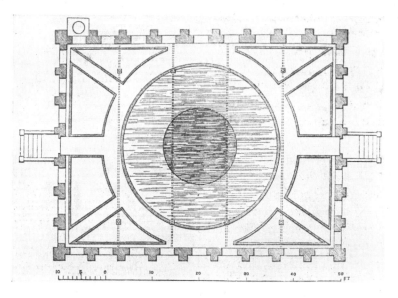

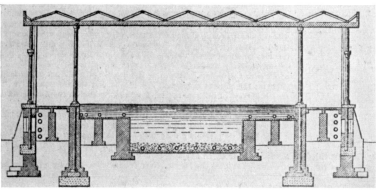

Victoria Regia water-lily house, Chatsworth, 1850, by Sir Joseph Paxton; plan and section.

more help; even while it deified *Les Grands Constructeurs* like Eiffel and Paxton, it failed to draw attention to the fact that Paxton was also a great and pioneering environmentalist, for whom glass walls and elaborate systems of heating and ventilating worked together to one end only, to make rare plants grow. Few of the buildings discussed in this book so far are comparable in terms of environmental ingenuity and performance with Paxton's house for the *Victoria Regia* lily at Chatsworth, completed in 1850, and it is probably true that an intelligent commercial glass-house operator today, judiciously metering temperature, moisture and carbon-dioxide levels in the atmosphere around his out-of-season chrysanthemums, has more environmental knowledge at his fingertips than most architects ever learn.

But if Le Corbusier, who knew about Paxton's Crystal Palace but, not, apparently, the Victoria Regia house, had wanted to have more physiological and environmental knowledge at his fingertips, it was probably more accessible to him than any other architect of his generation. His enquiring mind and his machine-age enthusiasms, his connections with industry, philosophers, medical men and *avant-garde* painters, gave him a range of intellectual contacts and sources of information far beyond that normal in an architect's professional life. The record of those contacts and sources is found throughout every one of the twenty-seven issues of the magazine *L'Esprit Nouveau* which he and Amedée Ozenfant edited between 1920 and 1926. But where is their record in his designs?

For instance, a sequence of issues (nos. 6, 7 and 8) carried an extensive summary coverage of the researches of Charles Blanc into human physiological responses to light, colour and form. The graphs that are reproduced look familiar still, exhibiting the characteristic curves of optical response to intensity and colour of light. But whereas one would expect to find them nowadays in the context of studies of lighting levels and environmental colour, Charles Blanc's interest, like that of Ozenfant and Le Corbusier, in undertaking the research (and theirs in publishing it) was to contribute to a 'rational' aesthetic of painting. No parallels with responses to coloured and illuminated spaces in buildings seem to be drawn or even implied . . . and this is the more astonishing when one considers the considerable emphasis given to human physiology in the same magazine by the indefatigable 'Dr Winter' who, issue by issue for some years, kept the readership informed of the latest records in athletics, and published in No. 15 an article entitled *Le Corps Nouveau*[2] with the slogan 'La physiologie est tout!'

[2] *Esprit Nouveau*, No. 15, p 1755.

But what Winter says, in detail, may give one a clue as to Le Corbusier's difficulties in putting all this information to work. Winter reveals himself a classicist, a fanatic of 'pagan' athleticism of the most routine kind. His rhetoric of *un corps nouveau riche d'un esprit nouveau*, is little more than rhetoric, in fact; an apotheosis of that mystique of the sportsman and athlete that was common in the generation that had grown up with Futurism, whence it had become a kind of cult among the *Ecole de Paris* artists

> The Euphoria of the athlete is pervading the whole world, and its empire will be immense. Painters, sculptors, poets must all submit, and a new artist will be born.[3]

[3] ibid.

No doubt the eurythmics taught by Le Corbusier's uncle, Albert Jeanneret, would be another manifestation of this new artistic euphoria, yet the general tone of the article suggests that, in the last analysis, Winter's *corps nouveau* would be old, in the same way as Le Corbusier's architecture is old; that is, classical.

> The body will re-appear, naked in the sunlight—cleansed, muscled, supple.[4]

[4] ibid.

The echo is almost too obvious—substitute the word 'architecture' for 'the body' and this could be Le Corbusier speaking. The new body was to be, it seems, the 'masterly, correct and magnificent play of muscles brought together in light.' This rehabilitated image of the Greek athlete, and the rehabilitated classical architecture to house it, are clearly aspects of the same vision, yet they seem to have no organic relationship. Standing upon a *terrain idéal*, not our common earth, each is a free-standing conception, absolute and unconditioned by the other, bathed by the same sun, but that sun does not shine through the windows of the ideal building into the eyes of the ideal body, nor does the breath of the ideal nose condense upon the ideal glass.

Yet Le Corbusier was neither totally unaware of environmental problems, nor indifferent to the human need to solve those that he recognised. Much of the *Manuel de l'Habitation* (which first appeared in the magazine, and was then reprinted in *Vers une Architecture*[5]) refers directly to environmental problems and methods for resolving them. Some of the propositions are little more than reflections of the Winter dogma, such as

> Demand a bathroom looking south, one of the largest rooms in the house . . .[6]

(though this would still be a great improvement on many French bathrooms one knows), but others are plainly sensible propositions even if laid down in consciously startling form, as:

> If you can, put the kitchen at the top of the house to avoid smells.
> Demand of your landlord that, instead of shades and drapes, he puts in concealed or diffused electric lighting . . .
> Demand a vacuum cleaner.
> Demand ventilators in the windows of every room.[7]

Now, it is an observable, indeed a conspicuous, fact that hardly any of these demands were satisfied in the houses that Le Corbusier designed in the years succeeding this pronouncement of 1921. He did not build any of them until the beginning of 1923, so none of these propositions are based upon recent constructional experience. They are, rather, based upon recent experience of living in rented accommodation in Paris ('Demand of your landlord . . . etc.') and addressed to other young intellectuals in a similar position to himself. It is less a manual of housing than a manifesto on converted property, rules for making oneself comfortable in the *quartier latin.*

For it is clear that many of these demands would not only be environmentally justifiable in such property, but could physically be answered. A standard glass ventilator not only should, but could be installed in the upper part of the traditionally tall Parisian window (and by now usually has been). Light fittings could indeed

[5] English edition, London, 1927, pp 114–115.

[6] ibid.

[7] ibid.

be concealed above cornices or recessed in the depth of a normally joisted ceiling construction. But they could not be so recessed or concealed anywhere on the flush walls or in the thin concrete floor slabs that Le Corbusier was about to start designing in late 1922. When the space above the ceiling was solid, and less deep than an average lamp-bulb with its socket-mounting, it was pointless to ask for it to be recessed, and it is technicalities of this sort that must account for the general tendency for lighting installations to be in full view in almost every one of the interiors he was designing while the rest of the world was just beginning to read *Vers une Architecture*, or the translations of it that followed from 1925 onwards. The most striking of all such cases, because of its effect on the photographic record, is that of the large living room of the Villa Cook of 1926. Splendidly lit by day with natural light entering from both ends, that entering from the south being given some masterly curved forms on which to play, it was illuminated at night by a single naked lamp-bulb mounted in the geometrical centre of its ceiling—a bulb so powerful that it burned a hole in the emulsion of the negative from which the illustration in Vol. I of the *Oeuvre Complète* was prepared, with the result that, as one can see without using a magnifying glass, the block-maker has had to redraw the bulb by hand!

The cause of the trouble was not merely that this was an exposed fitting, but that the very bulb itself was in clear view. Instances of this unsympathetic usage are far more common in Le Corbusier's work of the twenties than in that of the Bauhaus connection. Let it be said at once that he also used an inventive variety of glass and metal shades, most of them of industrial origin, but that the naked and exposed bulb was always an acceptable alternative in his eyes, almost to 1930. The usage is so persistent, and especially conspicuous in showpiece interiors where he was putting himself and his ideas on view, such as the highly programmatic *Pavillon de l'Esprit Nouveau* of 1925, that one is forced to suppose that this nudism of the light-source must also have been programmatic.

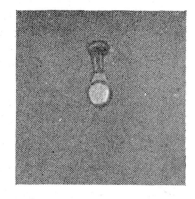

Villa Cook, Boulogne-sur-Seine, France, 1926, by Le Corbusier and Pierre Jeanneret; opposite: the living room; above: the lampbulb as re-drawn by the block-maker for the *Oeuvre Complète*.

148

As will appear from the quotation from Dr Winter, above, and from writings of the period *passim*, 'naked' was a good word at the time, and accorded well with the period's professed admiration for absolute honesty. Detailed confirmation that such concepts did influence the attitudes of Le Corbusier and his friends towards light sources can, in fact, be found in *L'Esprit Nouveau* and in an issue close in date to both the Pavilion and the Villa Cook. In No. 21, in an article entitled simply *Style Moderne*, but sub-titled '*Analyse des Formes et des Fonctions de l'Objet Usuel*' a correspondent named Yves Labasque laid down rules for objects of daily use in domestic interiors (many of them entirely sensible and laudable). In order to make one of the customary points against decoration, however, he selects a very telling example from the point of view of the present study:

> Useful objects must not dissimulate their functions; one form of decorative aberration is to lie about the nature and use of objects—Example; electric lamp disguised as oil lamp . . .[8]

[8] *Esprit Nouveau*, No. 21; no page number.

The words 'must not' (*ne doit pas*) are probably intended to have the force of a law of nature in the context of this argument, which is literally determinist in tone—the verb 'déterminer' appears in nearly every part and tense and in almost every other sentence. But if it is an offence against Natural Law to disguise an electric lamp, what constitutes an undissimulated one? Is a diffusing globe permissible, or is it an improper disguise because diffusing globes are found on oil lamps? Since such globes, in connection with oil or gas fuels for lamps, are commonly part of a glass funnel system that helps to control the flow of air needed for combustion, and since electric lamps need no such air, it may well have appeared to rationalising minds that diffusing globes were indeed an immoral dissimulation of the nature of the light being supplied.

If this had been the case with Le Corbusier in 1925, then it had certainly changed by 1930 and the design of the Villa Savoye, where indirect lighting is the rule, and the use of the living room ceiling as a major reflecting surface to distribute light throughout the

Villa Savoye, Poissy, France, 1930, by Le Corbusier and Pierre Jeanneret; the living room showing inverted lighting trough.

room from tubular lamps concealed above an inverted trough hung from the centre line of the ceiling, brings his artificial lighting techniques into a condition analogous to his intentions in day-lighting—'What use is a window, if not to light the walls?'. Now, both by day and by night, light both natural and artificial, comes less from a direct source than by reflection from an architectural sur-face. Unfortunately, the logical pursuit of his professed aims in natural daylighting had, by this point, arrived at a condition where it was defeating its own objectives.

The all-glass wall was that logical outcome, and much has been made, in an unconsidered way as often as not, of its environmental horrors—especially the tremendous solar heat gain due to green-house effects in summer, and the equally tremendous thermal losses in winter due to the almost non-existent insulating qualities of thin window-glass. Although Le Corbusier, as the apostle of the *pan de verre*, could not plead, as Wright could, a cheap-fuel economy and countervailing advantages, it should be remembered that he took over the fully glazed wall as a going tradition from Paris studio-house vernacular practice. Throughout most of the twenties, he used it only for studio rooms, or rooms of a studio type, and his clients were mostly art-lovers and others who already had experience of this kind of environment. They would therefore be prepared for, if not tolerant of, the concomitant expenditure on heating (which might not have been so very great, traditionally, because such houses were normally parts of close-knit urban terraces, with only their two shorter walls exposed, unlike Wright's deeply articulated and free-standing villas).

Furthermore, studio windows traditionally faced north, in order to receive the preferred invariant north-light around which the painting traditions of the school of Paris were built. All of Le Corbusier's early studio houses follow this rule of orientation as far as site-constraints permit, to within a few compass degrees (as anyone will discover if he tries to photograph their glass walls at the wrong time of day) and the solar heat gain through these windows is usually negligible. The sun does not significantly shine upon them, in spite of all the rhetoric—from Dr Winter, too—about the virtues of sunlight.

This situation lasted until about 1927, when the single house he designed for the Werkbund exhibition at Weissenhof, Stuttgart, no longer turned its two-storey studio-type window sternly to the north, but looks, instead, almost into the eye of the mid-afternoon sun—and it is difficult to see how it could do otherwise without either overlooking the back windows of the adjoining double house that he also designed, or totally missing the splendid views over the city of Stuttgart which lies almost due south. But, whatever the special excuses that can be advanced for this house, this seems to be the point where he abandons the common-sense orientation of his all-glass walls. The two big institutional buildings which, with the

Pavillon Suisse, Paris, France, 1931, by le Corbusier and Pierre Jeanneret: exterior showing main glazed façade facing south.

Villa Savoye, terminate and crown his work of the twenties—the Pavillon Suisse and the *Cité de Refuge*—both have their main glazed elevations facing within a few degrees of due south and take the full impact of the midday and afternoon sun. Both, in consequence, have presented serious solar heating problems.

The site of the *Cité de Refuge* may have left him little option about orientation, but what is so mystifying about the case of the Pavillon Suisse is that the site, when he moved on to it, was unencumbered, not overlooked by other buildings, and large enough to accept other possible groupings of the 'elements' of the plan, while the budget—whatever Le Corbusier may have said about it —was still liberal enough to permit some experiments in sound insulation, using suspended lead sheets in the hollow light-weight partitions between one student's room and the next. Ear-witness accounts seem to agree that this stratagem was not very effective . . . 'you can hear an electric razor three rooms away' is one assessment. But the historical point is that conscious provision for sound-insulation was actually made, for this shows that Le Corbusier was already having to confront the problems he had brought upon himself by trying to rationalise away the mass of the traditional masonry wall.

For he had argued himself into the position where he was faced in a most extreme form by the problems of the 'thin, tall and flimsy buildings' that had excited the concern of Konrad Meier thirty years before. The detailed process of this argument occupies much of his writings of the twenties, and has its origins in earlier writers, notably in Choisy's concept of frame-and-fill construction. Since the frame could be shown demonstrably to hold the building up, and the fill appeared to do nothing so useful, it was logical (if not sensible) to exalt the frame and to demote the fill to ever less substantial functions. By 1926, Le Corbusier had virtually abolished the physical presence of the filling membrane, if one is to believe the argument of his lecture at the Sorbonne. First he postulates his five 'Objective Elements of Discussion':

1. Architecture: to construct a shelter.
2. Shelter: to put a covering over walls.
3. Covering: to span an opening and leave clear space.
4. Light the shelter: to make windows.
5. Window: to span an opening.

and a few lines later he concludes from this

> But now a house can be built of a few reinforced concrete posts . . . leaving total voids in between . . . What good is it, I ask, to fill this space up again, when it has been given to me empty?[9]

In that epoch when *mince* must have been one of his highest terms of praise, all other materials seem to have been in his eyes

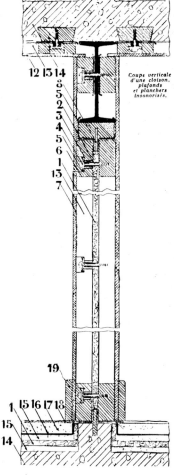

Coupe verticale d'une cloison, plafonds et planchers insonorisés.

Pavillon Suisse; above and opposite: sections through internal partitions.

[9] *Journal de Psychologie Normale et Pathologique*, No. 23, 1926, p 330.

poor substitutes for glass, his ideal of the de-materialised building skin, the minimum membrane between indoors and out. When he uses more obviously solid materials, as in the curved rubble wall of the Pavillon Suisse, it has the air of a surrealist quotation from an earlier culture, and where he uses lightweight but opaque cladding, as in the reconstructed-stone panelling of the end and rear walls of the dormitory block of the Pavillon Suisse it has the air of a grudging concession to human pudicity. Walls had been reduced to the level of a notional infill in the interstices of the frame, without mass or substance, and the advantages of mass and substance—thermal capacity, heat insulation, visual privacy, sound insulation, somewhere to drive a nail to hang a picture or cut a chase for conduitry—had all been abandoned.

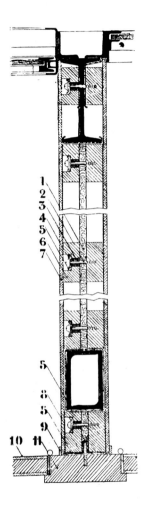

It is obvious that by 1930 he was becoming conscious of what he had done, what environmental qualities had been mislaid in his attempts to abolish the load-bearing wall. He was to discover, now, any number of good reasons 'to fill this space up again when it has been given to me empty.' To fill it up with suspended lead sheets for sound insulation, to fill it up with curtains to exclude sun and staring eyes, to fill it up with glass bricks for related reasons, to fill it up with solid masonry and other 'materials friendly to man', or to start thickening up the glass membrane with sun-breakers on the exterior, further layers of glass containing warmed air on the interior. In short, to replace additively, element by clip-on element, the performance factors that a massive wall had contained homogeneously and organically.

The most telling historical summary of this process of additive replacement is afforded by the Cité de Refuge. The multistorey glass wall of this immense hostel block faces a little west of south, and the nature of the site, on the Rue Cantagrel, gave him little option, probably, in fixing its orientation, or the general composition of the elements of the building, with the entrance, auditorium and other unique facilities in a cluster of geometrical solids at the foot of, and in front of, the glass cliff of the repetitive dormitory

Cité de Refuge, Paris, France, 1932, by le Corbusier and Pierre Jeanneret; left: the glass façade as originally built; opposite: with brise-soleil added.

accommodation. In the visual conception of the design there can be no doubt that this huge reflecting surface worked admirably as a backdrop to the cluster of auxiliaries assembled in sunlight, but as a cladding for the interior economy of the building its deficiencies due to Le Corbusier's obstinate environmental misapprehensions were doubly compounded by some shortcomings in the budget, and the idiotic manner in which the Préfecture de la Seine insisted on applying its town-planning regulations.

According to Le Corbusier's own account, he came to the task of designing the Cité de Refuge, already armed with what he considered to be the two master concepts of a new approach to environmental management: *la respiration exacte* and *le mur neutralisant*. The former concept meant, simply, controlled mechanical ventilation and is traced by Le Corbusier to the system employed by Gustave Lyon in the Salle Pleyel (the celebrated concert room, for which Lyon also devised the acoustic treatment), and the

neutralising wall is simply double glazing with hot or cold air circulating within the space between the two skins. He refers to this
as *notre invention*, and possibly with justification, since he had
used it once before in his career, while still in Switzerland, though
in a less extreme form. Both of these master concepts will need
further examination; their relationship to the Cité de Refuge he
sets out as follows:

> We had been looking for an opportunity—it came: the Salvation Army
> hostel *Cité de Refuge*. Six hundred poor souls, men and women, live
> there. We gave them freely the ineffable joy of full sunlight. A thousand
> square metres of glass wall lit every room, from floor to ceiling, from
> wall to wall . . . the glass was hermetically sealed, because warmed and
> filtered air circulates constantly inside, controlled by the heaters and
> fans.[10]

This sealed box of controlled ventilation and ineffable sunlight
opened triumphantly, because comfortably warm, in the bitter
December of 1933, as Le Corbusier proudly claims. And as he also

[10] *Quand les Cathédrales étaient
Blanches*, Paris, 1937, pp 25–26.

frankly admitted, it ran into serious trouble at the other solstice *au gros de l'été, à la pointe de chaleur.* The sealed glass wall with its southerly aspect made the interior an intolerable glasshouse; for reasons of economy it was a single skin, and not a *mur neutralisant*, and even if it had been, it would have made much difference, by Le Corbusier's reckoning, because the same budgetary restrictions meant that there was no cooling equipment in the ventilating system. In the upshot, the town planning authorities insisted on the fitting of openable *fenêtres d'illusion* whose environmental performance seems to have been less illusory than Le Corbusier liked to pretend, while he himself was driven, shortly after, to invent the external sunshade or *brise-soleil.* He did not, however, fit such a sunshade to the Cité de Refuge—that had to wait for another hand and a much later date.

As has been said, the invention of the sunshade or *brise-soleil* is an example of the process by which the advantages of the traditional massive wall were argued back one at a time. The transparent glass membrane, admitting the ineffable joys of sunlight, sufficient to stop rain blowing in and people falling out, could not exclude overdoses of the ineffable. An external egg-crate of vertical and horizontal shades could do this, however, while leaving the view almost uninterrupted—there can be no doubt that, however desperate its motivations, the *brise-soleil* is one of his most masterly inventions, one of the few last *structural* innovations in the field of environmental management that we have seen.

The other proposed solution to the solar heat-load problem, the neutralising wall, belongs, obviously enough, to the power-consuming tradition of environmental techniques, and is virtually Le Corbusier's first public recognition that such techniques were available. But it was not the first time the concept had passed through his mind, for that extraordinary last building of his Swiss career, the villa Schwob in la Chaux de Fonds (of which he was so proud until about 1923, and then suppressed for many years because of embarrassments about its style) had had a form of *mur neutralisant.* Its structural innovations as a pioneer (1915) concrete-frame design for a private house are well known, but it seems less known that the interstices of the frame were largely filled by double walling, the interval between the two panels being used for wiring and pipe runs. It also had double glazing for all the main windows, some of which were extremely large, one of them a two-storey affair, with heating pipes laid across the bottom between the two layers of glass, to obviate cold down-draughts due to the chilling of air against the window in winter.

The true neutralising wall, however, was to be a far more complex affair, both technically and in terms of the vast implications he supposed to stem from its employment.

> Every nation builds houses for its own climate. At this time of international interpenetration of scientific techniques, I propose: one single building for all nations and climates, the house with *respiration exacte.*
>
> I draw these floors, and I install the plant for *respiration exacte* . . . I make air at 18°C and at a humidity related to the state of the weather. A fan blows this air through judiciously disposed ducts, and diffusers have been created to prevent draughts. This regime of 18°C will be the arterial system, and I also supply a venous system, by means of a second fan which pulls in the same amount of air, and thus establishes a closed circuit. Exhaled and returned air goes back to the ventilating plant where chemical baths remove the carbon dioxide and it then goes on to be regenerated in an ozonifier, and into coolers if it has been over-heated.[11]

[11] *Précisions*, Paris, 1930, pp 64*ff.*

One realises that he really does mean an hermetically sealed system like that of a space-capsule or atomic submarine, not that either of these terms of comparison existed at the time. Arguing narrowly from first principles he has come up with another of his stunningly crude and absolute solutions, unreal and unworkable, to an entirely real problem. Willis Carrier, whose pragmatic approach will be discussed in the next chapter, had long before arrived at solutions far more subtle, economical, flexible and practical, of which Le Corbusier appears to have been totally ignorant—he seems never to have used the words 'air-conditioning' or their French equivalents, until after his visit to the USA in 1935–1936.

What would have rendered his absolutist solution even more uneconomical, was its wasteful duplication of plant, for we have not yet got to the *mur neutralisant* itself—the system described in the quotation above deals only with the ventilating air within the sealed box of neutralising walls; it is not intended to deal with thermal loads.

> I do not heat the building, nor the air. But an abundant flow of pure air circulates at 18°C to the measure of 80 litres per minute per person.
>
> And here is the second part of the operation.
>
> How, you ask, does your air . . . keep its temperature as it diffuses through the rooms, if it is forty degrees above or below zero outside?
>
> Reply, there are *murs neutralisants* (our invention) to stop the air at 18°C undergoing any external influence. These walls are envisaged in glass, stone, or mixed forms, consisting of a double membrane with a space of a few centimetres between them . . . a space that surrounds the building underneath, up the walls, over the roof terrace.
>
> Another thermal plant is installed for heating and cooling, two fans, one blowing, one sucking; another closed circuit.
>
> In the narrow space between the membranes is blown scorching hot air, if in Moscow, iced air if in Dakar. Result, we control things so that the surface of the interior membrane holds 18°C. And there you are![12]

[12] loc. cit.

And there you are indeed, all over the world, pegged to a standard temperature of eighteen centigrade whether you liked it or not.

> The buildings of Russia, Paris, Suez or Buenos Aires, the steamer crossing the Equator, will be hermetically closed. In winter warmed, in summer cooled, which means that pure controlled air at 18°C circulates within for ever.[13]

[13] loc. cit.

Rarely had his passion for 'le standard, l'invariant' been pushed to such pointless and impractical extremes, and never with such strong pretence of practicality, for Le Corbusier refers, in *La Ville Radieuse*, to practical tests on the *mur neutralisant* conducted in the Saint Gobain glass company's laboratories under the direction of Gustave Lyon himself, and cites laudatory passages from the Saint Gobain house-magazine *Verres et Glaces*. There is, however, a deception here, possibly a self-deception of the sort to which architects are prone at times. The laudatory passages

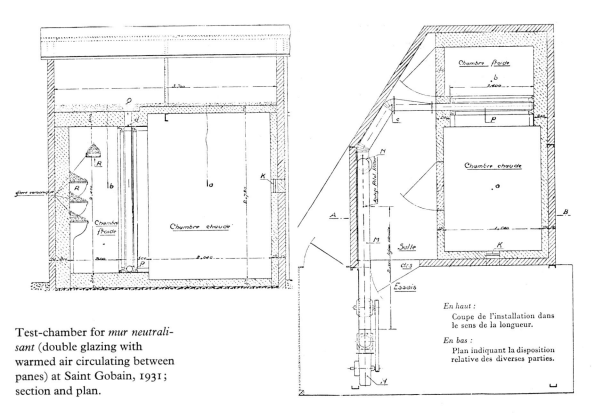

Test-chamber for *mur neutralisant* (double glazing with warmed air circulating between panes) at Saint Gobain, 1931; section and plan.

En haut :
   Coupe de l'installation dans le sens de la longueur.

En bas :
   Plan indiquant la disposition relative des diverses parties.

come from a general article about the intentions of Le Corbusier and Pierre Jeanneret in proposing the neutralising wall—and how could Saint Gobain find anything but praise for a proposition that involved using twice as much glass!—but not from the article which reports the conduct and results of the test. This is a rather different kind of text, a scrupulous and prosy account of how the tests were set up, the equipment used, the adjustments and measurements made. It would be an extremely difficult task to extract any obviously laudatory passages from this text, because there are none. The nearest thing to praise is

Warming the air between the panes increases the sensation of comfort.[14]

[14] *Verres et Glaces*, Aug./Sept. 1932, p 16. The whole document has special value for the light it throws on the aims and techniques of environmental research at the time.

and this is nowhere near emphatic enough to justify quotation by Le Corbusier in his own defence. What Saint Gobain technicians (headed, in fact, by J. le Braz) found were a number of interesting detail facts and unsensational improvements, and concluded that the use of hot air between double glazing needed a third layer of glass, trapping a second layer of (still) air, to make it a workable proposition!

But none of this in any way diminishes Le Corbusier's readiness to recognise the real thing when he encountered it, and to respond generously—his predilection for closed systems (in both the metaphorical and physical senses) did not prevent him having an open mind about the virtues of what Carrier had pragmatically devised in the US. In *Quand les Cathédrales étaient Blanches*, for instance, he relishes the interior arrangements of Radio City in phraseology as deft and efficient as the mechanical services of the building:

> A solemn temple, hung with sombre marbles, gleaming with clear mirrors framed in stainless steel. Silence, vast corridors and landings. Doors open, revealing silent lifts discharging clients. No windows anywhere, muted walls. Conditioned air everywhere, pure, dust-free, tempered . . .[15]

[15] *Quand les Cathédrales* . . . etc., p 46.

There is a fitting sense of occasion in this cool rhetoric—the ability to deliver conditioned air throughout a building of this size was a recent development, and justified in this case only by the imperative needs of ventilating sound-proofed broadcasting studios. The development process had been in train since the turn of the century —while European modern architects had been trying to devise a style that would 'civilise technology', US engineers had devised a technology that would make the modern style of architecture habitable by civilised human beings. In the process they had come within an ace of producing a workable alternative to buildings as the unique means of managing the environment, and had thus come within an ace of making architecture culturally obsolete, at least in the senses in which the word 'architecture' had been traditionally understood, the sense in which Le Corbusier had written *Vers une Architecture*.

Not that the inventors of air-conditioning knew this, or were interested in the outcome. Their aim was to make buildings as much more habitable as the commercial traffic would bear, and they had got so far with this purely *ad hoc* proposition that, by the time Le Corbusier had driven his aesthetic to the point where even he was convinced that it was necessary to invent air-conditioning to make it habitable, they could already have offered him a sophisticated plant that would deliver levels of performance beyond his conception. In achieving this, they had made the first major breakthrough since Edison's subdivision of the electric light. But this time the process was slower; the subdivision of conditioned air into room-sized supplies had been a long and tortuous process, but by 1929 it had been achieved, on an experimental basis at least (offices and boardroom of the Lyle Corporation, Newark, N.J.), and it was therefore possible conveniently to air-condition the kind of buildings that architects were commonly called upon to design.

*   *   *

But before turning to the history of air-conditioning, which now becomes necessary to the comprehension of further developments, there are two buildings of domestic scale which deserve discussion as substantial footnotes, at the very least, to this account of Le Corbusier. One is—in a very special sense—an unquestioned masterpiece in its own right: the house of Dr Dalsace, designed by Pierre Chareau and Bernard Bijvoet in the same space of time—1928–1931—as Le Corbusier was designing the large institutional works that have just been discussed. The name commonly given to the Dalsace House—*La Maison de Verre*—describes well enough its obvious environmental characteristic, the façade of glass bricks on its public elevation towards the rue Guillaume; lensed bricks

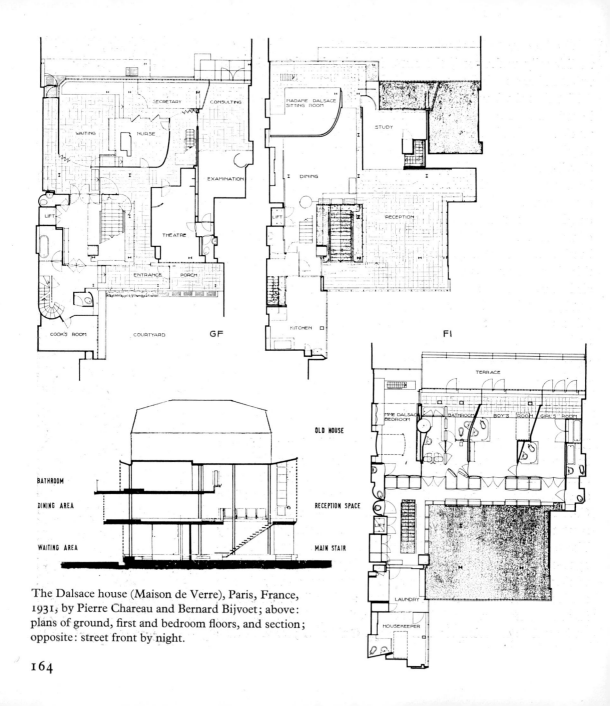

The Dalsace house (Maison de Verre), Paris, France, 1931, by Pierre Chareau and Bernard Bijvoet; above: plans of ground, first and bedroom floors, and section; opposite: street front by night.

Dalsace House; above and opposite: the living room.

which diffuse a controlled light throughout its main, balconied living-space.

However, the 'Machine Aesthetic' pretensions of this elaborate and introvertedly handsome interior are matched—and how rare this was at the time—by its mechanical performance. As Kenneth Frampton has written:

> ... throughout the house all electric and telephone wiring is conveyed vertically from level to level inside free-standing tubular metal standards that span from floor to ceiling. These tubes accommodate, in addition to the wiring, vertical control panels in which all outlets, switches, etc. are located ... As these thin vertical tubes carry power and light, so do the heavy floor-planes convey heat in the form of ducted warm air which is emitted from grilles above the floor surfaces.[16]

[16] *Arena* (Architectural Association Journal), April 1966, p 261. Frampton's account of the *Maison de Verre* in this issue of *Arena* is the only extensive description of the house in English.

These grilles are, in fact, recessed into a thickening of the floor plane which makes a kind of upstand kerb around its perimeter, and similar thickenings of the slab, but downstand on its underside, are also used to accommodate recessed reflectors for lamps. So here was one house in Le Corbusier's adopted city which, in the process of making a kind of continuous space sculpture of its mechanical services, managed to answer nearly all the environmental requirements of the *Manuel de l'Habitation*. But its Machine Aesthetic is not Le Corbusier's Machine aesthetic and it could never be mistaken for one of his works.

There is, however, one house which comes close to satisfying the *Manuel* and might be mistaken for his work in some views. The Aluminaire house, designed by two of his fellow-Swiss—Kocher and Frey—now stands on Long Island, but was originally devised for the celebrated Architectural League exhibition of modern houses in New York in 1931. Though not 'built like an aircraft, which is a little house that can fly and resist tempests' it was indeed of lightweight construction: steel angles, two-by-two studding, insulation board, plus corrugated aluminium cladding overall. Carried on slender pilotis, it has one *fenêtre en long*, Corbusier-style, one two-storey studio window, giving inevitably on a

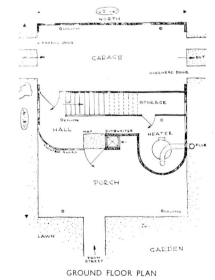

Aluminaire House, Long Island, N.Y., 1931, by Kocher and Frey; facing page: front elevation; left: two-storey studio window on side elevation; below: plans.

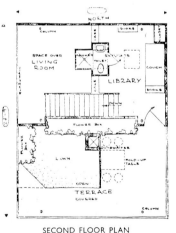

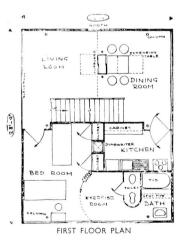

SECOND FLOOR PLAN

FIRST FLOOR PLAN

GROUND FLOOR PLAN

169

Corbusian two-storey living room roofed by a Corbu-type open roof-terrace. There are minimal bathrooms, recessed lighting tubes, translucent interior walls. All in all, it represents a telling collision between Corbusian architectural dogma, and the US technology needed to make that dogma come true. Telling because it works so much less well than the undogmatic approaches being made to that same technology by Rudolph Schindler and Richard Neutra (see chapter 10) over on the Pacific coast in the previous five or ten years.

Coming with a far more determined aesthetic (their work for the rest of the thirties shows far less variation in style than say Schindler's in the previous decade) they tried to force it on a technology to which it was not native. Lightweight US construction is well adapted to single-storey houses set directly on a brick basement, not to multi-storey affairs balanced on six pilotis, and its lack of inherent stiffness made it almost impossible to erect without elaborate temporary bracing (and permanent extra stiffeners). Internal acoustic privacy was minimal, especially for those who sat at the table permanently fixed between two downpipes, the one resonantly transmitting the effluent from the W.C. above, the other doing the same for the shower. In a normal US light-weight house, the spread nature of the plan on one or one-and-a-half floors can rarely produce such situations. The whole house is an eloquent demonstration of the differences between the ideal of the machine-aesthetic as preached by Le Corbusier, and the facts of machine technology as they existed on what had become, by then, their home ground, the USA.

9. Towards full control

As the progress of Le Corbusier's thinking shows, it would have been necessary to invent air-conditioning around 1930 had it not existed already. What makes the situation even more striking is that the development of the art of air-conditioning was itself reaching a point where its future growth seemed to demand a closer integration into the kind of building-design with which architects were normally concerned. For the history of air-conditioning is almost the classic example of a technology applied first in units of large capacity to industrial needs and to correct grossly deleterious atmospheric conditions, and then slowly sophisticated towards a condition where it could be subdivided and rendered subtle enough to handle domestic requirements.

The narrative of this process concerns, once again, the genial application of available scientific knowledge, or time-honoured rules of thumb, in piecemeal packets to piecemeal problems as they became apparent. But if this resembles the history of electric lighting in general outline, it is not dramatised by any single burst of concentrated systems-invention, such as Edison achieved, around 1880, nor is it ornamented by any personalities quite of Edison's quality. Willis Havilland Carrier has as good a right to be known as the father of his art, as Edison of his, but emerges from even the most eulogistic biographies as a man of more limited vision who, at least, began by evolving pragmatic solutions to *ad hoc* problems put in his way by other people. One might even, in an unsympathetic presentation, say that he did not recognise a problem until someone else offered him money to solve it. In his own words, 'I fish only for edible fish, and hunt only for edible game—even in the laboratory.'[1]

He seemed so content to solve problems as they were put to

[1] cited by Ingels, but already legendary.

him—often with startling ingenuity and depth of technical or intellectual resource—that one may doubt whether he had any general mental conception of the art he was founding until long after he had fathered it. The very words 'air-conditioning' are not his own, but were coined by his early competitor, Stuart W. Cramer, who used them more than once, in lectures and patent-documents, in 1904-1906. The Carrier Corporation, on the other hand, was still using phraseology like 'Man-made weather' as late as 1933, by which time the words 'air-conditioning' were general in the trade and were on the point of becoming part of common US usage—and had already appeared in the name of at least one of the numerous companies floated at various times around the personality and talents of Carrier himself.

Yet the phrase 'Man-made weather' is an admirable one, not only in describing the end product of the air-conditioning process, but because it also underlines the extent to which Carrier's mastery of the craft turned upon direct observation of the nature and performance of air as a component of outdoor weather. Thus his most crucial patent, dew-point control, for which application was filed in the *annus mirabilis* of this business, 1906, depended on a personal confrontation with the facts of fog, on a railroad station at Pittsburgh, late in 1902. According to Carrier's own account, recalled in old age, his response to air so laden with water droplets as to impede the sight, was:

> Here is air approximately 100% saturated with moisture. The temperature is low so, even though saturated, there is not much moisture. There could not be, at so low a temperature. Now, if I can saturate the air and control the temperature at saturation, I can get air with any amount of moisture I want in it.[2]

Such an observation cannot have failed to occur to others beside Carrier, once the mechanics of atmospheric humidity were understood, but by phrasing the matter in this way, he would almost automatically suggest a mechanism whereby that moisture could be controlled—to govern the absolute water vapour content of a

[2] Ingels, *Willis Carrier, Father of Air-Conditioning*, Garden City, 1952, p 15.

body of air by holding it, in the presence of excess water, at the temperature at which the maximum of water vapour it could be made to hold was the same as the amount desired, and then to remove the excess water droplets and restore the air to the temperature at which it was required to be circulated. This, obviously, meant regulating the temperature of the air twice, once to achieve the correct dew-point conditions required to regulate the total water-content, and then once more, to restore (usually to a higher temperature) the correct thermal content for circulation. Where Carrier put his observations of the fog to the most crafty use was in devising a method of achieving the dew-point temperature that was so brilliant and so paradoxical that it occurred to none of his contemporaries (there seem to have been no competing patents) and is still incomprehensible to many people today. His account of the Pittsburgh vision continues:

> I can do it, too (*scil.*, 'get air with any amount of moisture I want.'), by drawing the air through a fine spray of water to create actual fog. By controlling the water temperature I can control the temperature at saturation. When very moist air is desired, I'll heat the water. When very dry air is desired, that is, air with a small amount of moisture, I'll use cold water to get low-temperature saturation. The cold-water spray will actually be the condensing surface. I certainly will get rid of the rusting difficulties that occur when using steel coils for condensing vapour in air. Water won't rust.[3]

[3] ibid.

A knowledge of normal high-school physics will confirm the propriety of Carrier's method, but common-sense still boggles at the realisation that, for most of the air-conditioning year in most of the climates where air-conditioning is necessary, Carrier was proposing to dry air by pumping it full of water—and this, not as a bench-top trick at a Christmas demonstration lecture, but as a commercial proposition, twenty-four hours a day. It did not become practical at once, however; some years of trial and error with types and dispositions of spray-nozzles, and with baffle-systems to remove unwanted air-borne droplets of water from the saturated air were required. But, in the end, by this technique and a variety of automatic controls (which were not all of Carrier invention) the human race was at last armed with a workably sophisticated device for controlling the most elusive of environmental discomforts—parched or humid air.

But one must emphasise that the human race possessed this long-awaited device only in very large packets, applicable for reasons of bulky plant and crude ducting chiefly to industrial needs. At the time that Carrier began his industrial career with the Buffalo Forge Company in 1902, the large body of experiment and innovating installation then proceeding in a largely unco-ordinated manner throughout the American (and, indeed, world) ventilating and heating industries was oriented almost entirely towards improvements in factory environments, because there alone were the problems big enough, and profitable enough, to bring the manufacturers of plant and its users together in situations where the economic advantage to both sides were clear enough. In other words, air-conditioning was a way of losing less, or making more, money. With one or two freakish exceptions concerning supreme legislative bodies (the British House of Commons in 1838) or chief executives (the dying President Garfield in 1881) who were deemed worthy of environmental aids beyond those awarded to ordinary mortals, industrial needs dominated: refrigeration and ventilation in ships, regulated hot air for drying tea, bulk cooling in breweries, dust-laying in tobacco factories, control of mould growth on celluloid, fibre-humidity in weaving, ventilation in mines. Ogden Doremus might rhetorically enquire, 'If they can cool dead hogs in Chicago, why not live bulls and bears on the New York Exchange?', but until it could be shown that profits on 'Change were sagging, no-one was going to consider the proposition.[4]

In many of the purely industrial applications, of course, human comfort was a lively consideration wherever profitability depended on the efficiency of the work-force—thus, the laying of tobacco dust made it possible for operatives of cigar rolling machines to

[4] in historical fact, Professor Doremus's rhetorical question was to be answered within a mere eleven years of its utterance, for Alfred Wolff installed some form of cooling plant in the Stock Exchange in 1904.

see what they were doing, and thus make fewer mistakes; the ventilation of mines made it possible for miners to stay alive by breathing in locations and situations where there was profitable coal to be worked, but natural ventilation could never reach. Even the Larkin Building would probably have shown less care for controlled ventilation had the external atmosphere been tolerable by the standards of the time—indeed, it has been argued that the avoidance of soiling of documents and fouling of office machinery by airborne smuts was the Larkin Company's main motive for accepting a sealed building. Even in the roughly air-conditioned Royal Victoria Hospital, Belfast, it was dire medical need, rather than thought for human comfort, that dictated the use of Plenum ventilation, and all that that entrained architecturally.

There were, in practice, few situations where simple human comfort offered a profit margin proportionally large enough to make investment worth while, and large enough in absolute terms too, to make investment possible, given the plant then available. Hotel dining-rooms and ball-rooms came within this class, as did Pullman cars and—above all—theatres. The concentration of large audiences in places of entertainment—where they will normally expect to be made comfortable as part of the service for which they have paid—has always posed extreme environmental problems. The form of the buildings commonly employed, where 'crowding due to the presence of galleries' had the same effect as Professor Jacob had observed in Non-conformist chapels, and the need to make them reasonably proof against external noise and other distractions, produced a situation of congestion and enclosure where the heat from the bodies of the audience was more than sufficient to maintain normal room temperatures. Thus it became the custom of the trade during the period 1920–1950 when cinemas were normally full (or nearly so) from around mid-day to late evening, to turn off the heating altogether about two hours after opening, except in very severe winters.

In warmer, Southern climates, the body heat load commonly became an embarrassment or even a hazard. The prevalence of fainting in the audiences certainly had more than purely dramatic causes, and the use of the fan was often as much an environmental necessity as an aid to flirtatious communication. With such a heat load, the chemical vitiation of the air became an even greater burden, but it would have been bad enough without the thermal hazard—some of the nineteenth-century's most spectacular concentrations of carbon dioxide were recorded in the pits of theatres. Nineteenth-century environmental engineers had made a start on these problems long before air-conditioning was even contemplated, of course—large public buildings with auditoria, like the Free Trade Hall, Manchester, or council chambers, as in Leeds Town Hall, were often provided with large thermal syphon extract ducts, powered by braziers or heating coils at their bottoms. In cases like Cuthbert Broderick's design at Leeds, these ducts could emerge above cornice level in bulk large enough to rival the intentional features of 'art architecture' and demand equally artistic detailing as a consequence.

But the availability of large-capacity fans toward the end of the century brought these hazards in sight of solution. Professor Jacob, as usual, gives a reliable survey of the state of the art at the time of his writing, and draws particular attention to two cases:

> The arrangements for the heating and ventilation of the Vienna Opera House are singularly complete. They were designed by Dr Bohm, the medical director of the Hospital Rudolfstiftung. There are two fans, one for propulsion, the other for exhaust. The air is heated by steam coils, and is admitted by the floor and through the risers of the seats. Each gallery and compartment, including the stage, has its own independent supply and means of heating . . . Air is admitted to a basement chamber, into which, in summer, sprays of water are introduced; it is then driven over the steam piping and on into a mixing chamber . . .
> Very similar arrangements are found in the Metropolitan Opera House, New York; but there is but one fan, and that is used on the 'plenum' system . . . to avoid draughts from the doors, which are so usual in theatres ventilated on the exhaust principle.[5]

[5] Jacob, *Notes* . . . etc., p 93.

Jacob also cites the case of the Madison Square Theatre in New York, as do other writers, because its Sturtevant fan-system, from 1880 onwards, had provision for blocks of ice to be stood in the intakes to cool the air, and could consume up to four tons of ice per night in summer.

Such cooling techniques could be capricious, of course; according to ambient circumstances, the input air might pick up moisture from the ice by evaporation, or lose it by condensation on the ice surface. Though the probability would normally be that these effects would have the right tendency—that hot dry air would pick up humidity, and hot humid air would, with luck, shed some—the system was not reliable enough to compensate for its cumbersome bulk, messy operation, impossibility of automatic control and constant demand for labour. Air-conditioning looked a more attractive proposition on all of these counts, and was bound to come in as soon as it was mechanically practicable. There appears to be some room for argument about which was the first of such 'theatre comfort jobs' but Margaret Ingels in her life of Carrier, awards the palm to Graumann's Metropolitan in Los Angeles, a Carrier installation with Carbondale refrigerating plant, of 1922.

The Graumann's installation, and the numerous other theatre and cinema comfort jobs which followed, all effectively reversed the ventilating proposition discussed in the quotation from Jacob, above. Whereas schemes such as Bohm's at the Vienna Opera had tended to use the space under the ramped seating as a distribution volume for the input air, which entered the auditorium under the seats, the new comfort jobs reversed the flow, bringing air in through diffusers overhead at low velocity, whence it settled in a cooled blanket gently over the whole auditorium, to be extracted through grilles in the risers under the seats. Given the fact that in most auditoria, cooling is a far greater problem than heating, and that this arrangement gives the preferred 'cool-head/warm feet' stratification, overhead input and under-seat extract is now almost a world standard. Most commercial theatres also developed, at an early date, the habit of deliberately spilling some of their conditioned air out of the foyer on to the side-walk, thus offering tangible proof that it was, indeed, 'cooler inside.'

The movie industry thus introduced the general public to the improved atmospheric environment, as well as the improved luminous environment which will be discussed in the next chapter. But could any members of that public enjoy that same improved environment at home, or even at work? Well before the end of the twenties it was clear that anyone who could reduce air-conditioning to an office-block scale, let alone a domestic one, had a bright commercial future. Traditionally, the earliest fully air-conditioned office block is taken to be the Milam Building in San Antonio, Texas, of 1928; architect, George Willis, and engineer, M. L. Diver. In spite of its uninspiring exterior, it was an innovating building on many counts—for instance it was among the first concrete-framed skyscrapers and, at twenty-one storeys, the tallest multi-storey concrete framed structure in the world at the time.

Its air-conditioning method was simple in general conception, though complex in application.[6] A common refrigerating source in the basement supplied, firstly, an air conditioning plant for the main public rooms on the lower floors, and secondly, a set of standardised smaller plants for the standardised office floors. These sets of machinery were distributed at the rate of one to every two floors, near enough, throughout the height of the block, and were located between the toilets and elevators at the back of the floor-plan: each set supplied conditioned air to two floors through ductwork in the furred spaces above the ceilings of the central corridors, and the corridors themselves served as the return ducts, exit grilles from the rooms being provided in the doors. This effected a reasonable and economic compromise between the unavoidable necessity of working with fairly large units of plant, and subdivision of their output without consuming too much rentable floor space with large vertical ductwork.

[6] there is a useful description of the building in *Heating, Piping and Air-Conditioning*, July 1927, pp 173ff.

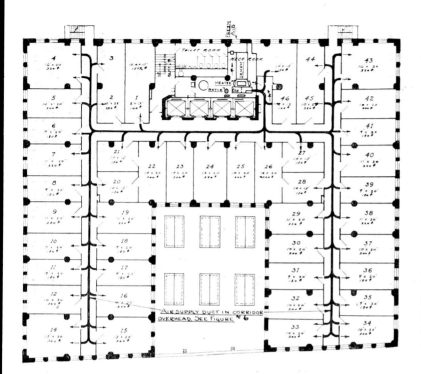

Milam Building, San Antonio, Texas, 1928, by George Willis; facing page: exterior view; left: plan of typical floor and below, section of standard duct and corridor arrangement.

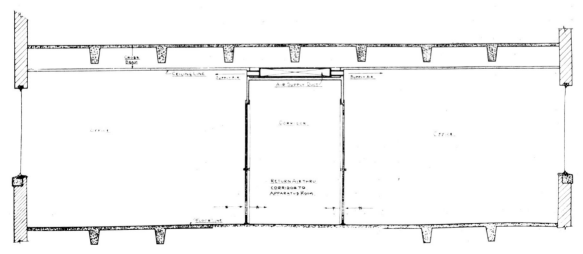

In such situations where commercial practicability was both the initial motivation and the ultimate veto, the consumption of floor-space by duct-work was a life-or-death consideration, since even the comforts of air-conditioning were rarely attractive enough for the rental to be elevated to the point where the loss of square-footage was offset. Carrier's solution—and ultimately everybody else's—was to distribute filtered and moisture-controlled air at high velocity through small diameter ducts, and heat it or cool it at the point of delivery under the windows of the offices by means of pipe-coils warmed or chilled by water supplied on a separate network. Also at the point of delivery, the unwelcome side-effects (noise, draught, etc.) of high-velocity distribution were tamed by using injector nozzle systems to make the conditioned air draw considerably more than its own bulk of room air through the casing of the unit, the mixture of new and locally recirculating air emerging quietly and at unobjectionable speed as a curtain in front of the window glass.

The embryo of this concept, and some of its essential parts, already existed in the board-room installation at the Lyle Corporation offices in 1929, to which reference was made in the last chapter, but it was not yet a sufficiently workable proposition to be used by Carrier in the Philadelphia Savings Fund building of 1932 (see next chapter). The full 'Conduit Weathermaster' installation as a standard kit did not exist until 1937, but something very like it can already be seen in the illustrations to an article, *Preliminary Planning for Air-Conditioning in the Design of Modern Buildings*, by two Carrier employees, Realto Cherne and Chester Nelson, which appeared in *Architectural Record* in 1934. But the importance of the article as an historical marker goes well beyond this point: the illustration shows, unmistakably, an office block divided up into small room-units, not a large single industrial or theatrical volume; the text says, 'This discussion will be limited to air-conditioning for comfort . . .',[7] and the whole represents the earliest ascertainable occasion on which air-conditioning appears

[7] *Architectural Record*, June 1934, pp 538*ff.*

to have been discussed at the level of the kind of conventional professional wisdom that is embodied in architectural check-lists.

The emphasis in air-conditioning had clearly changed; but for the tide of recession and economic collapse that swept North America, the use of air-conditioning in large, multi-celled buildings would probably have become established before the thirties were over. As it turned out, however, the rate of progress was slow, and the expensive installations at Radio City that so impressed Le Corbusier were without significant rivals after 1932. In some ways, it may be argued, this delay may have benefited both architecture and air-conditioning. With the Second World War following even before economic activity was fully recovered from the Slump, a total of more than a decade was amputated from the expected growth of air-conditioning. Real progress was not fully resumed until after the end of the forties, by which time the mechanical possibilities for office-block air-conditioning had been reinforced by a new technical aid in the field of lighting, and a new set of aesthetic preferences in the design of building envelopes.

The innovation in lighting was the fluorescent tube, which, with its relation the gas-discharge tube, had existed as a potentiality since the beginning of the century. Claude, in France, and Moore, in England, had produced workable discharge tubes at an early date—Moore tubes had been used to outline the façade of the West-End Cinema in London in 1913, and Claude's favoured discharge-gas—Neon—had added a new word to the language before 1930. But it took from Edison's 1896 experiments with fluorescent bulbs until the simultaneous announcement by Westinghouse and GEC of their 'Lumiline' tube in the summer of 1938, to get the fluorescent tube into the catalogue and onto the market.

To the world at large, and in the minds of architects, the fluorescent tube was essentially a post-War innovation, and was prized primarily for its economy of current and its lack of concentrated glare. But even in the first Lumiline announcement, the relevance of the fluorescent tube's diminished heat output to problems of air-conditioning was mentioned. And in any case, the use of fluorescent lighting was soon to generate new glare problems when it was employed in even and continuous grids over office ceilings as a source of PSALI in areas too remote from the perimeter windows of the block to receive much usable daylight. But such use of PSALI (Permanent Supplementary Artificial Lighting of Interiors) at the core of very deep floor-plans could never have come about without the neat confluence of the potentials of air-conditioning and fluorescent light. The heat output of enough incandescents to give a tolerable level of illumination for paper-work would have been more than any ventilating system could economically have swept out. But with a diminished heat output, air-conditioning could cope economically, and once this was possible it also became possible to make a long overdue rationalisation of the standard US office tower's plan-form.

Traditionally, it had always exhibited a notch or re-entrant in the back (Europeans will probably know it best from the plan of Holabird and Roche's Marquette building which is so often illustrated in accounts of the Chicago School) and this re-entrant served to bring light (and ventilating air) to the centre of the block, including its ancillaries, such as toilets and elevator shafts, as well as rentable office areas. But a plan bitten into in this way was more difficult to subdivide and contained more awkward corners that were difficult to let, than would the plain rectangle of what was to be called the 'full-floor' type of plan, with its ancillaries islanded in the centre—a possibility that existed, profitably, only with air-conditioning and low-heat lighting. Given these, however, it was calculated by a Chicago real-estate man, George R. Bailey, that

Full-floor development can be produced, complete with air-conditioning, fluorescent lighting and acoustic ceilings, for only about 8% more than a standard floor (i.e., with notchback) without air-conditioning and with only ordinary lighting.[8]

His calculus was timely—not only was the clear, well-serviced rectangular floor plan attractive enough for its rents to absorb that

[8] *Heating, Piping* . . . etc., September 1949, p 72.

extra eight per cent, but architects had by now more or less unanimously decided that their post-War skyscraper dreams were going to be realised in a starkly rectangular aesthetic. Both the United Nations building and Lever House were in design and construction at the time Bailey's results were published, and though both were prestige buildings which, for differing reasons, could support 'uneconomical' standards of servicing, the innumerable rectangular glass slabs which appeared in their imitation soon showed that such a format, and its necessary standard of servicing was not at all uneconomic—or, at any rate, not unprofitable. Le Corbusier's vision of the Cartesian glass prism of the slab skyscraper, and Carrier's practical technology for solving any environmental problem that offered an honest dollar had met, literally, in the UN building, and the face of the urban world has been altered.

But, even at that date, the interior of the domestic dwelling was still virtually unmarked by these upheavals of environmental technology—air conditioning was just beginning to find its way into the home in 1950. The story of its arrival had been a long and—for the trade—frustrating one. The ultimate historical reasons probably lie in the peculiarity of the industry itself, and the kind of men who led it. Men like Carrier, even when employed by commercial concerns, usually worked for companies that produced only part of the total kit needed for air-conditioning—the fans, or the refrigerating plant—and saw air-conditioning primarily as a means of promoting the parent company's sales. Almost like independent consultants, they assembled the total plant from the wares of several manufacturers, often by separate competitive tenderings. Nobody in the earlier stages appears to have manufactured, or even offered to sell, a complete installation as a pre-assembled package. The elements of the kit were distributed, according to private lores and mysteries of consultancy and subcontracting, within the interstices of the building-structure, and the layman therefore had difficulty in identifying air-conditioning plant as a commodity or recognisable service such as he might be able to install in some convenient space in his own home.

What made this situation the more frustrating for the trade's visionaries and opinion-makers was that such convenient spaces existed almost universally throughout North America, in the common house-basement, and that those spaces already contained testimony of the wide-distribution of the skills needed to install air-conditioning, in the shape of the ductwork taking hot air from the furnace to the various rooms—indeed, these ducts would often have served well enough for conditioned air as they stood. Even a small opening into this promising market could, like office-block installations, have helped the industry round the awkward corners of the Slump. A rousing editorial in the Chicago magazine, *The Aerologist* during the summer of 1931 coined the splendid slogan: *Wanted, an Air-Conditioning Flivver!*, and called for

> . . . an air-conditioning unit for the home, efficient, moderately priced and relatively fool-proof . . .
>
> Its production on a quantity basis by modern manufacturing methods would soon make air-conditioning more of a necessity than the radio or even the automobile, and its acceptance in the home would soon force its general adoption on a grander scale in practically every other building and conveyance used by man.[9]

Effectively, history was to run in the opposite direction—effectively 'every other building and conveyance' would be air-conditioned before there would emerge a domestic air-conditioner as ubiquitous as the family flivver, and the process was ultimately to take almost two full decades from the publication of the *Aerologist* editorial. In the mean time, there were to be isolated and expensive installations in luxury homes, some, indeed before 1930, such as those in the Chicago area by the redoubtable Samuel R. Lewis. The General Electric Company installed an experimental room cooler for evaluation in Carrier's own house in 1929. There followed a flurry of interest in room cooler units, though most of them, unlike the example just cited, were not self-contained but serviced by a refrigeration plant somewhere else, usually in the basement.

[9] *Aerologist*, August 1931, front-cover editorial. Published in Chicago in the twenties and thirties, *Aerologist* was one of the few publications concerned with the general atmospheric good of the human race, not with narrowly technical or hygienic aspects of the topic.

Carrier, by now involved willy-nilly in manufacturing, *via* his Standard Products division, had an 'Atmospheric Cabinet' room cooler on the market by 1932, but this was still too bulky a block of equipment to recommend itself as domestic furniture. Most of the central-station units intended to service the house through ductwork from the basement were even more cumbersome. Even in the flattering light of carefully air-brushed advertisements of the mid-thirties, they are seen still to be *ad hoc* assemblies of the needed units, mounted—sometimes—on a common base and grudgingly wrapped in characteristic examples of industrial stylists' case-work of the period—though the Trane Company appear to have despised even this minimal concession to domesticity and continued to glory in a Boilermakers' Aesthetic of pipe-elbows and exposed valves. And they were bulky, commonly occupying a near cube of about six feet by six feet by six feet, heavy to match and costing more than $2000 in some cases. These figures alone would have made them an unattractive domestic proposition for the mass market, even had they been more foolproof and more nearly self-regulating than they were. Ten years after the *Aerologist* had formulated the need, the Flivver had still to materialise, and the US went to war without any domestic airconditioning to return to[10].

But there was not long to wait after the War. As usual, hostilities had stimulated the rate of invention and technological development, to the point of precipitating some minor technical revolutions, not all of which had any relevance to air-conditioning as they stood, but all tending to point the way towards a radical miniaturisation of the equipment involved. The cumulative effect of miniaturisation and other improvements was to be suddenly sensational around 1950. Writing with the slightly dazed air of someone who cannot quite believe his eyes, Arthur Carson observed in 1954:

> Research that started in 1946 hit the production line with its discoveries in 1951, when mass-produced home air-conditioning units appeared on

[10] Professor Condit has suggested to me that the slowdown of the progress (and miniaturisation) of air-conditioning in the 1930's may not have been as total as I have suggested, because of the continuing installation and improvement of railway air-conditioning in the US. By 1936 all lounge, dining and sleeping cars on major long distance trains in the US were air-conditioned.

AIR CONDITIONERS

AIR CONDITIONER YEAR 'ROUND
(Floor Model)

AIR CONDITIONER

AIR CONDITIONER YEAR 'ROUND
(Suspended Model)

COMFORT COOLER

From low cost comfort coolers to dual purpose year 'round air conditioners McQuay emphasizes efficiency—durability. Designed for quiet, dependable operation, McQuay air conditioners offer the benefits of controlled air comfort for small shop and factory alike. Air conditioning

units cool or heat, humidify or dehumidify, filter and circulate air for all-season use.

Available for water or direct expansion cooling, steam or water heating, McQuay has a big selection of models and capacities to meet your needs.

McQuay Company, packaged air conditioners, 1948.

the market in every shape and form. In 1952 dealers sold out $250,000,000 worth of equipment and had to turn away 100,000 customers. That year there were only 20 companies in the field; now there are more than 70, with the original 20 multiplying their 1952 output by 400–500%.[11]

[11] Carson, *How to Keep Cool*, New York, 1954, p 56.

Carson perhaps may be faulted on his starting dates, since some of this research appears to have hit the production line earlier than he suggests—both the McQuay Company, and General Electric, seem to have had the kind of air-conditioning pack he is describing in their catalogues in 1948, and there are some doubtful cases as early as 1946. One says 'doubtful cases' because the kind of air-conditioning unit that Carson is discussing is quite specific but not at all what the editorial in the *Aerologist* had anticipated. What had finally wrought the revolution and brought in the air conditioning Flivver was not a central station system servicing the house through ducts, it was not a room-cooler with a remote refrigeration plant, it was not a compound unit like Carrier's Weathermasters. It was a simple, self-contained box, needing connection only to an electrical outlet; it could usually be lifted by one man, or two if unusually large; its bulk might not be more than two or three cubic feet; and it provided full and complete air conditioning, with the possible exception of winter humidifying which is rarely needed anyhow. What makes some of the early units 'doubtful cases' is that it is not now certain how full a conditioning service they offered, but what Carson is talking about is a self-contained unit that can be installed in a hole in the wall or an opened window, plugged in to the electrical main, and can deliver genuine air-conditioning.

That is all the installation it gets in many cases, rested on a window-sill and the sash closed on top of it—many models nowadays come ready with flaps or fitted plates to block off the residual width of the window opening. Although domestic air-conditioning of the sort Carrier had envisaged, from a central station using ductwork in common with the winter heating, has also proliferated, it has done so in the wake of the self-contained window unit, which has finally made air-conditioning comprehensible as domestic equipment comparable with the cooker, the refrigerator and the television set—a neat box with control knobs and a mains connection. However one regards this device, it is a portent in the history of architecture.

Firstly, by providing almost total control of the atmospheric variables of temperature, humidity and purity, it has demolished almost all the environmental constraints on design that have survived the other great breakthrough, electric lighting. For anyone who is prepared to foot the consequent bill for power consumed, it is now possible to live in almost any type or form of house one likes to name in any region of the world that takes the fancy. Given this convenient climatic package one may live under low ceilings in the humid tropics, behind thin walls in the arctic and under uninsulated roofs in the desert. All precepts for climatic compensation through structure and form are rendered obsolete—though as James Marston Fitch (and others) have hastened to point out, any consideration of economy in the use of air-conditioning brings the

Lafayette Park apartments, Detroit, Mich., 1961, by Mies van der Rohe; left: exterior of blocks; below: specially developed air-conditioner package to fit under window.

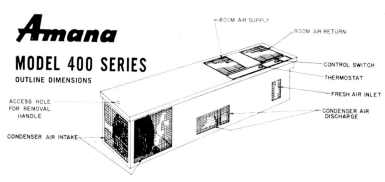

Amana

MODEL 400 SERIES
OUTLINE DIMENSIONS

ROOM AIR SUPPLY

ROOM AIR RETURN

CONTROL SWITCH

THERMOSTAT

FRESH AIR INLET

ACCESS HOLE
FOR REMOVAL
HANDLE

CONDENSER AIR
DISCHARGE

CONDENSER AIR INTAKE

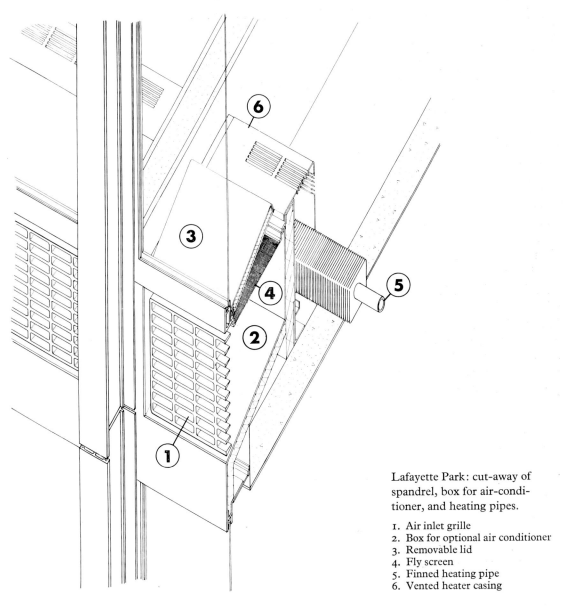

Lafayette Park: cut-away of
spandrel, box for air-conditioner, and heating pipes.

1. Air inlet grille
2. Box for optional air conditioner
3. Removable lid
4. Fly screen
5. Finned heating pipe
6. Vented heater casing

time-honoured usages of local tradition back with even greater force.

Nevertheless, the possibility of absolute variety and infinite choice of building form is now with us—and as so often happens with infinite choices, has led to almost perfect homogenisation of what is chosen. In the United States, air-conditioning has now made the established lightweight tract-developers' house habitable throughout the nation, and since this is the house that the US building industry is geared to produce above all others, it is now endemic from Maine to California, Seattle to Miami, from the Rockies to the bayous. Not without reason; the normative American house that Catherine Beecher saw evolving out of the normative US way of life that grew from the increasingly mechanised farming of the middle-west, probably answers to the understood aims and usages of the people who inhabit it quite as well as any localised vernacular house of the Old World suits its tradition-bound inhabitants. The house type was already widespread and still spreading long before air conditioning came along to wipe out its surviving deficiencies, and much of its adaptation to increasingly specialised climatic conditions was the work of tenants and owner-occupiers who fitted packaged air-conditioning to their own homes with their own hands in their spare time.

For this, secondly, may prove to be the most portentous aspect of air-conditioning, that, in the domestic package, it offers the most sophisticated device for environmental management that mankind has ever possessed, in a form that needs little skill to install, and even less to operate. Any normally intelligent householder can install one with normal household tools and many have done so. As far as the little houses of suburbia are concerned, this poses no great visual or architectural problems—the evergreens have already grown up in front of the units, and they are not seen. But in the apartment blocks of cities, such installations can bring the environmental improvements of the householder into direct conflict with the visual intentions of the architect. For every act of

intelligently permissive vision, like that of Mies van der Rohe at the Lafayette Park apartments in Detroit, where a characteristically well-detailed under-window box offers the householder choice between controlled natural ventilation and the installation of an optional air-conditioner purpose-designed to drop into the box, there are too many designs where conflict seems inevitable.

A conspicuous case is that of the Kips Bay Apartments in New York, designed by I. M. Pei and Associates. Since this is a scheme of 1959–61 the absence of provision for air-conditioning is not altogether easy to understand, though the determination of the tenants to fit it, especially to rooms on the south faces of the blocks, is entirely comprehensible. In order to preserve Pei's façade patterns, the managing company is reputed to have insisted that

Kips Bay Apartments, New York, 1961, by I. M. Pei Associates; southern façade showing privately installed air-conditioners.

tenants should install their conditioner packages, not in the spand-rels under the windows, but back inside the room and connect them to the outside air with a flexible duct. Since, however, the environmental performance of almost every domestic air-con-ditioner depends on its being able to dump its surplus heat and moisture directly into the outside atmosphere (as any one will know who has been wept on by air-conditioners in the streets of New York) this hopeful proposition has not proven successful, and most of the tenants have their units back in the spandrels under the window—in a random pattern all over the façade that is not unattractive.

Although many architect-designed buildings are now beginning to make their peace with the seemingly inevitable eruption of room-conditioners on their façades, few have set out to exploit the neat visual detailing of their intake grilles, nor the convenience for interchangeability of their easy installation and removal. A rightly-noticed exception to this indifference is seen in the terrace of row-houses in the Old Town area of Chicago, designed by Harry Weese in 1963. There, the conditioner grilles form rather delicate visual accents on the outer faces of projecting cupboard units that occupy several bays of the frame on the back of the terrace, while the non-structural nature of these projecting cupboard-backs

Row-houses, Eugenie Lane, Chicago, Ill., 1963, by Harry Weese; above and facing page: rear elevation showing air-conditioners in projecting cup-board units.

should make them—as the *Architectural Review*[12] observed— 'relatively simple to alter when the air-conditioners have been overtaken by the normal processes of technological obsolescence.'

Unfortunately, this terrace on Eugenie Lane is notable chiefly because it is an exception to the general failure of architects to make provision for this piece of equipment which is rapidly becoming as normal as the kitchen sink. And unfortunately again, this failure of provision does not, in residential work, produce even the accidentally picturesque disorder that often arises from the air-conditioning of standard US single-storey commercial structures, where below eaves level, one is confronted with an orgy of eclectic modernistic details often reaching extremes of uninhibited fantasy and above the eaves the air-conditioning plant is a free composition of geometrical solids of the sort that used to be the common stock in trade of the Purists and Functionalists of the twenties. One seems to see ghosts of the restrained and abstract International Style hovering above the exuberance of the current pop-art version of the American high-life which Oud discovered in the work of Frank Lloyd Wright. Except that what is above the eaves does not represent a conscious attempt at an idealised Machine Aesthetic, but is the outward form of the kit environmentally needed to make the high-life of supermarket-America possible.

[12] *Architectural Review*, May 1964, p 311.

10. Concealed power

The calculations of George Bailey on the economics of the slab skyscraper, cited in the previous chapter, bring together three environmental aids—acoustic tiling, air-conditioning, and fluorescent lighting—that were to add up into one of the key inventions of recent architecture (though no architect invented it): the suspended ceiling. One should add immediately that ceilings have been suspended almost as long as there has been architecture, but whereas the traditional reason for suspending a ceiling was to close the top of a room volume, the sense in which the word is employed nowadays in trade literature and office-conversation, implies opening up the top of the room volume to admit environmental power over its entire area. Paradoxically, this opening-up is commonly considered as an act of concealment; the ceiling, in spite of its general perforation, is seen as a way of concealing the fact that the upper part of the room volume is occupied by ducts, conduits and service adits generally.

The earliest documented use of this type of concealment that has come to hand, is A. M. Feldman's ventilation installation[1] in the Kuhn and Loeb bank in 1906, mentioned in chapter 4. The problem was to cool the banking hall (the building still stands, at the junction of William and Pine Streets, near Wall Street, but none of the installation appears to have survived) which was on the corner of the block, with exposure to both streets. The only available surface on which the plant could reasonably be mounted was the flat roof of an extension at the back of the block. Between this and the banking hall was the manager's office, lit through a glass roof. Ductwork taken through the upper part of this room would be conspicuously seen in back-lit silhouette, quite apart from any other environmental disadvantages that might accrue,

[1] the only full account is in a paper read to the American Society of Mechanical Engineers in 1909 by Feldman himself (New York Public Library, Bound Pamphlets, vew pv6, No. 12), but the installation is mentioned in the Ingels chronology.

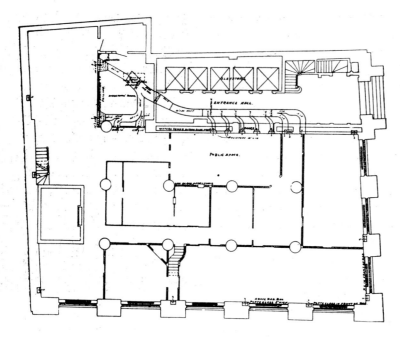

Kuhn and Loeb Bank, New York, 1906, ventilation installation by A. M. Feldman; plan showing overhead duct-work.

and Feldman dealt with the problem by suspending a ceiling of obscured glass at the level of the internal cornice, thus retaining at least some of the room's natural lighting but concealing the ductwork.

Even though the installation did not service the room in which this ceiling was suspended, the intention of concealment was much the same as that animating most later suspenders of ceilings. The fact that these concealments take place normally within the formal rules of an aesthetic (the International Style) conspicuously given to honest exhibition of structures and services (see chapters 7 and 8, *passim*) will suggest that the story of the emergence of the suspended ceiling as we know it now must be somewhat confused, both in its narrative sequence and in the intentions of those involved. Though this is true to some extent, and represents the all too common twentieth-century situation of technology pushing architectural concepts beyond their original intentions, one should point to at least one group of architects who were less pushed by technological innovation because less committed to a pseudo-technological 'Machine Aesthetic', and the less confused in their intentions for the same reason.

History, as written so far, still tends to concentrate attention on the European practitioners of the International Style as the main driving and innovating force in modern architecture. But the International Style, however important in Europe and the Europe-oriented Eastern states of the USA, was never the whole of modern architecture, and one can indicate two other streams of modernism, ultimately fused, which prefigure later developments more accurately, even if they had no very direct influence upon them.

One such stream, obviously, involves and descends from, Frank Lloyd Wright and his scattered discipleship. A recurring interest in luminous forms, containing concealed light-sources, can be seen in much of his work, and the concealed lighting of the Robie house has already been discussed in chapter 6. From that period, to the complex system of glass tubing, backlit, between the mushroom caps of the columns of the vast office-space of the Johnson Wax Company in 1936, one can find innumerable ideas for concealed and diffused lighting in his projects and built work. But even in 1936 he still was not yet seized of the idea of the true suspended ceiling, since what came from the upper surface of the Johnson Wax office volume was purely light; the supply of ventilating air was provided by blowers concealed with some difficulty in the thickness of the balcony floor-slabs.

A more consequential line of influence, though difficult as yet to trace in detail, may prove to be the one that descends through Wright's more marginal followers. Thus, when Albert Chase McArthur designed the Arizona Biltmore hotel in Phoenix in 1927–1928

Mr Wright came out to Arizona and all the technical details for the use

Johnson Wax Company offices, Racine, Wis., 1936, by Frank Lloyd Wright; interior of typing pool.

of the concrete block type of construction were worked out under his direction.[2]

Unfortunately, this small act of friendly consultancy backfired somewhat, and McArthur was to find that the consultant's great name was taking away his own reputation, and had to write to *Architectural Record* in 1941 a letter re-asserting his authorship. The mis-attribution is understandable however, for the design contains many obviously Wrightian features, and also makes a logical development from the concrete-block work of the Master in his California houses. That development was, most obviously, the use of a lighting-block coherent and modular to the structural blocks and their decoration:

[2] *Architectural Record*, July 1929, pp 19ff.

> the lighting system of the Biltmore hotel likewise was considered as an integral part of the architectural design, the architect deliberately seeking to avoid the afterthought effect of the usual methods of illumination. In the main, the lighting scheme consists of substituting for the concrete blocks a number of pressed glass blocks set in frames of sheet copper and flush with the walls.[3]

[3] op. cit., p 23.

In practice, the effects could be more visually adventurous than this quotation might suggest, because the lighting bricks could occur in corner situations with two surfaces exposed, or as major decorative features in, say, the re-entrant between a structural column and a ceiling beam. Furthermore, there was a standard ventilating brick in the same decorative idiom and modular sizes, perforated with a Wrightian pattern of slots, which is used throughout the building, in conjunction with the hollow spaces inside the walls, to promote natural ventilation (apparently with success). But while it is gratifying to find an environmental device so neatly detailed into the structural system, there were to be few more of consequence that worked without mechanical power. The lighting solution devised for the Biltmore, on the other hand, clearly had advantages that could be taken up, and something like it can be seen on the large structural columns that rose through the interiors that Jacques Peters designed for the Bullock's-Wilshire store in Los Angeles in 1929.

Arizona Biltmore Hotel, Phoenix, Ariz., 1928, by Albert Chase McArthur; interior showing perforated blocks for ventilation.

Arizona Biltmore: the main lobby with recessed lighting panels.

There is obviously no need to hunt for direct influences from the Biltmore in such an instance. Wright himself was in and out of the State of California, members of his family were established or working there, his former *homme-de-charge*, Rudolph Schindler, was working there—and was certainly in touch with Peters by 1930 if not earlier.[4] But if this was a connection with Wright, it was a very special connection, which also included Richard Neutra and J. R. Davidson, and had involved roots in Europe, from which most of its members came. In the work they did in California,

[4] Peters was one of those who, with other members of the Neutra/Schindler connection, including Wright, took part in the exhibition in 1931 which led to the final rupture between Schindler and Wright.

particularly in lighting, there are some striking similarities of intention (and realisation) with the work that Mendelsohn was doing in Berlin at the time. This was hardly surprising—Neutra, who had come to the US largely at Schindler's insistence, had shared an office briefly with Mendelsohn before leaving Europe. J. R. Davidson had worked in Berlin, and was well-acquainted with Mendelsohn's work, before he left for California in 1923. Mendelsohn was in the US in 1924, and was introduced to Wright by Neutra, shortly before Neutra too left for the West Coast.

Thus, one can demonstrate a continuity of personal contacts between Berlin designers and Californians down to the beginning of 1925, but there is precious little else to connect the two bodies of work; the memories of those who established themselves in California in the early twenties are eloquent on the resultant total break of contact with Europe, and their increasing ignorance of what was being built even in the eastern US. Few international publications reached the Pacific coast (which puts paid to such historian's wish-thoughts as the proposition that Schindler could have known projects by Mart Stam that were published in 1925), and magazines outside California seem to have been entirely ignorant of the highly original architecture being done there until *Architectural Record* began to publish the work of Schindler and Davidson in 1929.

So it would be foolhardy indeed to speak of any kind of unified California/Berlin school, but there remain some historiographical

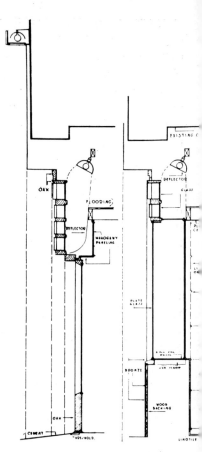

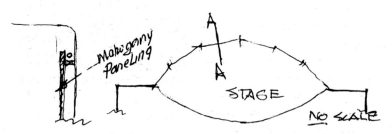

Left: back-lighting in piano-showroom, Berlin, 1923, and; above: concealed lighting for shop-fronts, Los Angeles, Cal., 1925, both by J. R. Davidson.

conveniences in taking Berlin and California together in the twenties, rather as Chicago and California can be taken together before 1910—especially as Davidson seems to have been an innovator in lighting in both places. Before he left Berlin he designed a piano showroom with lights concealed behind projected panelling on the wall, shining upwards into the cove of the ceiling, and in his own offices in Los Angeles designed in 1925, he appears to have been one of the first to use regular commercial glass bricks, back-lit, as part of an exterior scheme, as well as specially designed fittings in the office itself.

Mendelsohn's lighting practice seems to have derived little from his US visit—and probably did not need to. The recessed lights above glass panels under the canopy at the entrance to the Berliner Tageblatt offices date from as early as 1922, two years before the American trip. If Neutra brought anything with him from his Berlin experiences, it is not easy to identify. Nor do any of these designers seem to have borrowed anything much from such obviously available influences as the drawings in *Frühlicht* or the work of Hans Poelzig, whose work in the Grosse Schauspielhaus abounds in concealed and coved lighting, but the use of largely reflected light is quite different from what Mendelsohn, for instance, set out to do with his flat bands of back-lit glass set flush with the flat surfaces of his architecture.

What is clear, in Mendelsohn's Herpich store of 1924, for instance, is that the intentions of the Berlin-California 'school' were to exploit the left spaces and hollows of normal construction to provide light without making its sources visible. In the main selling space of the store, light comes from luminous wall panels above the merchandise racks, from coved lights in a projecting shelf above these racks, and from a broad band of back-lit glass in the ceiling, following the perimeter of the room plan at a distance of about fifteen inches from the walls. The spaces to provide these lights followed almost naturally from normal shopfitters' structural methods, and are thus directly comparable to

Rudolf Mosse offices, Berlin, 1923, Eric Mendelsohn; recessed lighting under canopy.

Grosse Schauspielhaus, Berlin, 1920, by Hans Poelzig; lighting column in foyer.

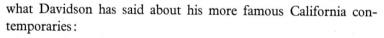

what Davidson has said about his more famous California contemporaries:

> ... both used flushed built-in lighting between beams or posts with semi-obscure glass panels.[5]

which is a similar exploitation of normal US house-building techniques. But Mendelsohn was to come even closer to the modern concept of a 'lumenated' ceiling in 1927–1928, in the interior of

Herpich store, Berlin, 1924, by Eric Mendelsohn; above: exterior by night; left: sales area with recessed lighting.

[5] letter from J. R. Davidson to the author.

Universum Cinema, Berlin, 1928, by Eric Mendelsohn; interior.

the Universum (Luxor Palast) cinema in Berlin.[6] There, the luminous strips amount to almost two-thirds of the ceiling area, and are an integral part of the pattern of sweeping Borax stream-line curves which constitutes the main motif of the interior decoration—and was to father so many good, bad and insufferable movie-house interiors in the ensuing twenty years.

It is tempting, at this point, to enquire whether the movie-industry may not have played a more direct part in this development of improved modern lighting in Los Angeles and Berlin. Both were major centres of movie-making during the period, both were backed by advanced electrical industries, both supported a liberated and permissive attitude on the part of architectural clients, even if this led to eccentric historical fantasies as often as

[6] Mendelsohn was clearly excited about the ventilation provisions too; see his poem on the Universum, cited by Dennis Sharp in *Modern Architecture and Expressionism*, London, 1966, pp 125–126.

it led to advanced modern architecture. Yet, in spite of the expanding technology of flood-, spot- and colour-lighting that the film-studios were demanding, and in spite of the increasing skill of designers and producers in using these technologies to create and manipulate virtual spaces and other illusions, there seems—as at the Bauhaus in the same years—to have been little cross-fertilisation. If anything, according to Davidson, the architects had to push the manufacturers, rather than their being inspired by the new products in the manufacturers' catalogues. One blockage may well have been that studio lighting equipment was too big and powerful to be even considered in the context of domestic lighting, and would certainly produce more heat than domestic ventilation would normally sweep out. So one finds Davidson employing perfectly ordinary filament bulbs in reflectors to back-light his glass bricks.

Where the influence of the performing arts can be seen most clearly in the work of the Berlin designers, is probably in the idea of bars and strips of luminosity within a rectangular picture-format—more likely a stage-set than a frame of film—which they applied to the façades of shops and movie houses; eg., the bars of light across the façade of Mendelsohn's Herpich store, made by flooding the spandrel panels from coved lights above the cornice of the floor below. Where the influence can be seen in Los Angeles is most probably in the free and open intellectual climate that Hollywood bred in those years of its most uncontrollable success. If this produced tendencies to the excessively flamboyant, sensational and luxurious, it also made room, intellectually, for some modest experimentation with inexpensive technologies.

In Rudolph Schindler's Lovell beach-house, a work contemporary in design with Rietveld's Schröder house, and built with the same materials though without the disguise of abstract idealism that Rietveld found necessary to apply, but of otherwise comparable architectural intentions and importance, the sources of light are not the kind of mechanistic sculptures or pseudo-indus-

Lovell beach-house, Newport Beach, Cal., 1926, by Rudolph Schindler; interior of living-room, showing 'light-tower' by window, and 'light-ladder' on wall at left.

trial fittings that Rietveld employed. They consist, for instance, of lamps concealed behind plain wooden baffles to throw their light across adjoining wall-surfaces, behind panels of obscured curved glass between structural joists, or low-power bulbs between stacks of close-packed wooden 'shelving' that ran up the walls in strips about a foot wide. Movable lights were simply-built pieces of furniture, 'light-towers' of alternating bands of light and dark. The same pattern repeats, more or less, in the corner glazing of the main windows of the same room—a congruence of natural and artificial lighting that would be impossible to parallel in the work of his more famous European contemporaries, and would probably never have occurred to them as a worthwhile objective in design.

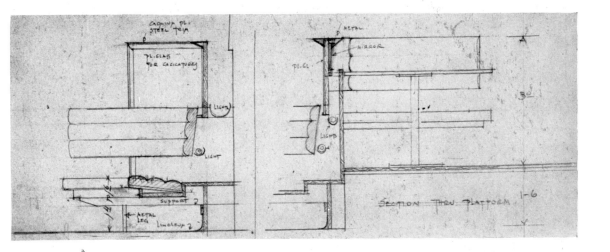

Similar felicities of integration, equally simply achieved, can be found in several of Schindler's domestic designs of the twenties, but another vein of adventure in luminous modelling of architectural form can be seen in works such as his interiors for Sardi's restaurant on Hollywood Boulevard in 1932. The interest here lay (and only briefly, for it was burned out and rebuilt within a couple of years) in the close integration of the lighting with the seating arrangements, which were in the form of semi-circular booths or banquettes along the wall, with light coming from a continuous trough below the cornice, from light fittings behind the seats, and from backlit glass panels between one booth and the next.

This, though striking in its own right as a deployment of light to enhance the quality of environment, does not have much to do with the invention of the suspended ceiling, but there is a design of a few years earlier by Richard Neutra which does seem to connect. When he, in turn, came under the patronage of the Lovell family, and built for them the famous house in the Hollywood hills in 1927, he not only executed such machine age *jeux-d'esprit* as using Ford headlamp reflectors in recessed fittings, but he also

Sardi's Restaurant, Hollywood, Cal., 1932, by Rudolph Schindler; drawings for concealed lighting behind seats.

devised for the long library a continuous hanging light-shelf practically fifty feet long, two feet below the structural ceiling and projecting some eighteen inches from the wall. Powered by thirty low-wattage colour-corrected 'daylight' bulbs, it threw light upwards by direct radiation on to the ceiling, and diffusedly downwards through the obscured glass forming the bottom of the shelf. This seems to be the first interior that any apologist has ever claimed for the International Style to be lit in so humane and intelligent a manner, but its contrast with admired contemporary European works (this was the year of Weissenhof with its bare bulbs and grim globes) is almost shocking.

American writers frequently single out Neutra's Lovell House as one of the key points for the beginning of the American contribution to the International Style (ironically missing the continuity of

Lovell House (Health House), Los Angeles, Cal., 1927, by Richard Neutra; suspended lighting in library.

the California school to which it belongs) and in this context it has been associated (by William Jordy, notably) with the first International Style building on the other side of the continent, the Philadelphia Savings Fund Society's office-tower of 1932. Such a classification misses the point of Howe and Lescaze's design by almost exactly the same proportion as it does Neutra's, for the technical and above all, environmental achievements of both buildings dwarf their stylistic consequence. Outwardly, PSFS is a rectangular office slab, probably the first, standing on a podium of shops and public rooms for banking, and backed by a subsidiary tower for elevators and services, etc. It thus gives Howe and Lescaze a good claim to have pioneered the characteristic format of Lever House, or—with its externalised structures and articulated service-tower—the Inland Steel Building in Chicago.

But besides these two famous works by Skidmore, Owings and Merrill it also anticipates another, their Harris Trust block of 1957 in Chicago. The much noticed 'missing' floors half way up this block are anticipated by a 'different' floor half way up PSFS, and for the same reason, the economies of installing the main environmental plant in the centre of the slab. For PSFS was a fully air-conditioned office tower, the second in the US (and thus the world) after the Milam Building (see previous chapter). The presence of its Carrier conditioning plant at twentieth floor level is signalled by a changed pattern of fenestration, and by air intakes. In fact, PSFS had a double plant, one serving the large volumes of public space in the podium, the other handling the more standardised loads in the office floors. But non-standardised variations of load in the office areas due to external climate could also be handled. The central plant on the twentieth floor had separate fan-systems for the east and west, and could thus compensate for shifting solar-heating loads.

The main distribution of conditioned air from this central plant was upwards and downwards through vertical ducts situated in the dead-spaces behind the service lifts, the central division ensuring

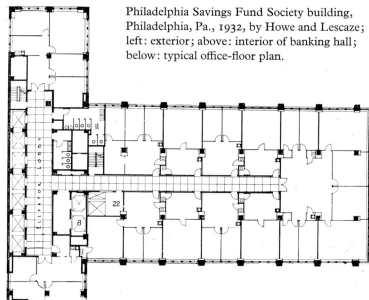

Philadelphia Savings Fund Society building, Philadelphia, Pa., 1932, by Howe and Lescaze; left: exterior; above: interior of banking hall; below: typical office-floor plan.

that no part of the vertical ducting acquired the uneconomic cross-sectional dimensions that would arise with a central plant at the top or bottom of the slab—no more than half the total volume of air was moving vertically at any one level. After this, however, the distribution was more conventional, pairs of horizontal ducts under the ceiling of each floor distributing the air outwards at the line of the outer face of the internal structural columns—that is, about a third of the way from the central corridor to the outer wall on either side of the slab. It was thus not strictly a corridor type system like that of the Milam, even though return air passed through louvres in the doors to the central corridor. Although the overhead location of the distributor ducts would almost automatically suggest a suspended ceiling nowadays, nothing of the sort seems to have been attempted at PSFS, and conventional light fittings are seen dangling in the early photographs.

The situation was different in the banking halls and public spaces, however. The main banking hall was lit by low fittings on the counters, and by indirect light thrown up on the ceiling from lamps concealed above dropped panels around the heads of the columns. Inlet and extract grilles for the air-conditioning were also inset in these columns, the ducts being concealed in massive furred spaces that swelled the visible section of the columns far beyond structural necessity. In other areas, such as the entrance lobby, furred spaces in the ceilings also served to house ducts, from which the air entered the lobby through annular diffusers around recessed lamp-fittings—again an anticipation of much later practice.[7]

How much extra heat load was imposed on the air-handling system by such usages as this is not clear, though it seems that filament lamps generally throughout the building caused thermal problems, and most of the lighting in all areas has been extensively altered. However, the point which is historically relevant here is that in these combined diffuser/lamp types of installation, architects and engineers were working together to exploit the lost

Harris Trust and Savings Bank Building, Chicago, Ill., 1957, by Skidmore Owings and Merrill; night view showing service floors (dark) half-way up slab.

[7] The deliberate use of extract air to remove the heat of light at source, however, seems to have waited until 1956, when Robert T. Dorsey installed a return-air troffer experimentally in his office at Nela Park (as narrated in his paper to the I.E.S. conference in Cambridge, 1968).

Philadelphia Savings Fund Society building; lobby, combinded lamp and ventilating diffuser at right.

volumes of the ceilings and beginning to treat the ceiling-surface as a multi-purpose membrane of concealed power—another contribution to the evolving concept of the suspended ceiling. And if the final emergence of that concept was still two dècades away when PSFS was completed, it is interesting to note that just as the suspended ceiling was to emerge finally, around 1950, there was a flurry of revived interest in PSFS—a long article by Frederick Gutheim.[8] for instance, full of praise—as if Howe and Lescaze's

[8] *Architectural Record*, October 1949, pp 88*ff*.

towcr had at last been recognised as the true ancestor of the architecture that was about to be.

Of that architecture—the glass slabs of the fifties—the suspended ceiling was an integral part, and its history as a marketable product in manufacturers' catalogues goes back to the period of PSFS, when a mixed batch of problems and inventions set the process going. One problem was fireproof construction, commonly achieved through the use of concrete floor slabs. As in Le Corbusier's case, this deprived architects of the useful dead spaces in traditional floor/ceiling construction in which to stow all sorts of useful equipment. Neither the casting of conduits in position, nor the cutting of chases overhead were attractive propositions, in terms of labour costs or otherwise, and could do little for bulky services like air-handling. Such things had to be slung under the slab and, in buildings of any pretensions beyond strict utility, concealed by conventional plastered ceilings.

Again, the steel industry, looking for new outlets in the economic difficulties of the early thirties, was also beginning to interest itself in fireproof floor structures, but from an opposite approach, employing open-web lightweight joists with light steel decking and screeded floor laid on top. The Rivet Grip Company had such a system in its advertisements as early as 1930, and it was soon seen that service runs could conveniently be threaded through, and carried upon, the members of the open truss, and a ceiling could then be attached directly to the bottom of the web. Alternatively, the Mellon Institute in Pittsburgh put forward in 1932 a floor of structural steel decking with screed laid on top, and a full kit of parts to hang from its underside, including duct-runs and a suspended ceiling. A new use for the hollow middle layer of such sandwiches was put forward in 1936 by Rivet Grip—to use the lost volume itself as a distributing plenum for conditioned air without ducting, and another, closely related system appeared almost at once, Burgess Acousti-Vent.

As the name of the Burgess system implies, this is the point

Corrugated steel floor/ceiling system, 1932, developed by the Mellon Research Institute, Pittsburgh.

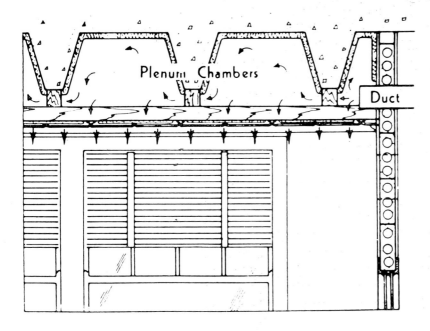

Plenum Chambers

Duct

Perforated Acousti-vent ceiling
system, 1936, developed by
Burgess Laboratories.

where the second element in George Bailey's calculus enters the
story, as acoustic tiles join with air-conditioning. Both systems
proposed to use the new product, perforated acoustic tile, to :—

(a) form the ceiling surface
(b) deaden the reverberation of sound
(c) form a continuous diffusing outlet for ventilating air.

We are therefore dealing with a genuine multi-purpose power-
membrane. Other power-membrane concepts appeared in the same
period because panel heating, which had been around in various
embryo forms for some decades, now became a practicable and
discussable idea. Practicable, because small bore heating pipe
techniques made it possible to turn a wall, floor or ceiling into a
surface radiating heat all over, by burying a suitable net of copper
pipes in the plaster-work or screed; and discussable because the

method was still so new that it threw up problems and advantages that the architectural profession had hurriedly to explain to itself and understand—notably the thermal storage capacities of such heated slabs, largely forgotten in lightweight American building.

However, the acoustic tile brought with it another kind of architectural consequence that is still effective today, almost forty years after its appearance. Unlike plaster, which may be spread over a ceiling of any dimension and shape, acoustic tile comes in discrete rectangular units, answering to a limited range of standard dimensions. It thus defines the effective module of the ceiling, encourages the use of other components of the same module and rectangular format in order to avoid wasteful cutting of tiles; it also fixes the pattern of hangers and ribs which will support the tiles, and thus constrains the disposition of whatever else is to go above the visible ceiling. As other parts of the suspended ceiling kit began to answer to the tile module, a dimensional inertia was built up, which resists variations of the tile sizes, and constrains the dimensions of any new technologies—such as the troffers for fluorescent tubes—that are added to the kit.[9] Sets of standard dimensions are thus created outside any intentional systems of preference or modular co-ordination, and affect other buildings in which acoustic tile is not employed. Thus the dimensional grid of the "Schools Construction System Development" (SCSD) building-system developed by Ezra Ehrenkrantz and his team at Stanford University in the early 1960's ostensibly rests upon the need to accommodate standard fluorescent tubes within the cells of the underside of the deeply-trussed roof structure which contains all the environmental equipment needed to service the rooms below. But the sizes of those tubes had been conditioned by the ceiling grids that accommodated earlier models, and so on—a history of accidentally determined sequences of dimensional standards reaching back into the early thirties.

The tyranny of the tile format was to become almost absolute, so

[9] the tube lengths were deliberately adapted to the module of the standard square 12-inch tile by a team at General Electric under Walter Sturrock, according to a note given me by Professor Buford L. Pickens.

De Laveaga Elementary School, Santa Cruz, Cal., 1966, by Leefe and Ehrenkrantz; above: exterior; below: diagram of the essential parts of the SCSD system. External walling is not part of the system and remains at the discretion of the architect.

1. Mixing boxes for air from conditioner on roof
2. Rigid distribution ducts
3. Flexible distribution ducts
4. Ceiling outlets
5. Lighting system
6. Roof space acting as return air plenum

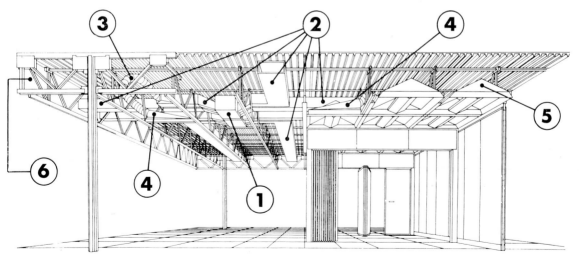

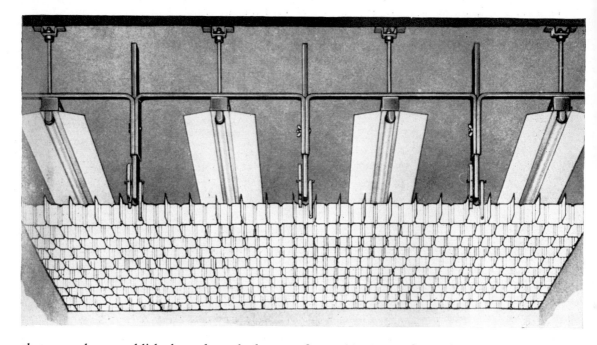

that even long-established products had to conform—'At last, the *square* Anemostat!' confessed advertisements in 1949, as the well-known and standardised line of circular diffusers finally came out in a square model. By that date however, the final steps toward the true suspended ceiling had been taken—from 1947 onwards, standard kits of parts existed (Benjamin Skyglo, Skyline Louverall) from which could be assembled systems of translucent, louvered or egg-crated suspended ceilings with suitable provision for the downward diffusion of conditioned air, either through specialised outlets or (in the louvered versions) over the whole ceiling. Having to be transparent to light and/or air, such systems had little acoustic value, and their ancestor, the acoustic tile, will often be found to have migrated back to the place where it started, stuck to the underside of the fireproof floor-slab, leaving only its module at ceiling level.

Suspended ceiling system, 1947; Skyline Louverall.

Thus the concept originally roughed out by A. M. Feldman at the beginning of the century became an industrial product by 1950, that is to say, it became possible to buy a servant-space off the peg, and hang it in the upper part of your room. Like many significant developments in recent architecture, it passed virtually without comment in the professional literature, until it was firmly established enough to form a subject of common complaint:

> The suspended ceiling, says architect Walter Netsch, is the soft under-belly of US building!
>
> His comment reflects the dissatisfaction of many architects and engineers who are far from content with the way today's conventional suspended ceiling looks and how it works.

reported *Architectural Forum* in 1963,[10] having taken hardly any editorial notice whatsoever of the growth of the concept and product during the previous thirty years. Netsch had some right to offer an opinion, however; he was a major designer for an office— Skidmore, Owings and Merrill—that had done a good deal to make architectural sense of the suspended ceiling concept by that date. But as far back as 1948, advertisers had been able to see quite clearly what the situation really involved. Copy-writers for Robertson Q-Deck, to take a conspicuous example, scornfully compared traditional fireproof slab-floors to tombstones, and contrasted them with their own system of steel decking, hung ductwork and suspended ceiling membranes, which they termed the 'vital arteries' of the building.

[10] *Architectural Forum*, May 1963, p 140.

Because the history of the suspended ceiling had been so little observed, and its parts were so inextricably integrated by 1950, the earliest serious commentators upon it could read from cause to effect in either direction, but because architectural habits of thought were so structure-oriented, the situation was still commonly seen as a distortion of structure by services. Thus L. W. Elliott commenting on Saarinen's Technical Center for General Motors properly observed that

> ... the equipping of most modern American buildings with compre-

hensive mechanical services has stimulated the development of suspended ceilings to contain ducts, pipe-runs, and service outlets . . .[11]

and adds an advantage that would recommend itself to pre-fabrication-obsessed English technocrats,

> . . . this method, as well as providing a high degree of noise-absorption, can also serve to establish a constant floor-height thus permitting the use of standard partition units.[12]

but then goes on to take what might seem an odd view of what is ultimately at stake:

> In the case of the office-building for the General Motors research centre, designed by Eero Saarinen, the systems of mechanical services are so extensive that the depth needed to house them is sufficient to allow for a built-up welded truss two foot six inches deep.[13]

But if this seems a strange way of expressing a matter which (if later Saarinen office gossip is to be trusted) was not reasoned out in that way at all, but a truss deep enough to clear-span the space was found deep enough for the services, the Saarinen office does seem to have made a somewhat oblique approach to the curtain-wall/suspended-ceiling combination. Though it is easy to assume nowadays that the two things belong together almost of historical necessity, the Drake University laboratories project of Saarinen, Swanson and Saarinen, still, in 1947, combines curtain walling with duct-work in the furred space of the corridor ceiling only, extracting directly from the labs through grilles in the corridor wall, above the door-line. Other buildings of the same vintage, by normal commercial architects, used analogous methods—the Universal Pictures building on Park Avenue, supposedly New York's first fully air-conditioned office block, had a central spine-duct fed by its own fan room on each floor, and the location of the duct was normally in the corridor, blowing into the offices directly through the corridor walls.

But, in praising the installations in the GM office block as the first 'integration of structure and services' worth serious dis-

[11] *Architectural Review*, April 1953, p 252.

[12] ibid.   [13] ibid.

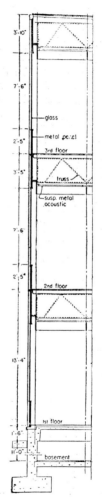

cussion, Elliott was undoubtedly right, it is a handsome and well-engineered scheme, and his article gains authority and relevance from the fact that it also discusses two of the most consequential buildings in this context ever built—the United Nations Headquarters and Lever House. The UN building would be consequential anyhow, simply as the symbol of the parliament of the world, but its secretariat tower has architectural and environmental significance beyond that. Whatever the niceties of the distribution of international credit for the design, the original conception is unmistakably Le Corbusier's. Here, in New York, he accomplished his dream of creating a great glass tower in an urban setting, and

General Motors Technical center, Warren, Mich., 1950, by Saarinen, Swanson and Saarinen; opposite: part section of structural frame; above: interior of drawing office, showing variability of ceiling system.

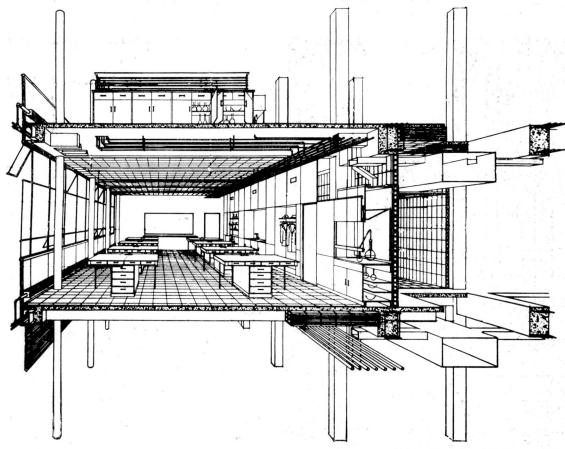

Drake University Laboratories (project), 1947, Saarinen, Swanson and Saarinen; cross section of complete structural and servicing ceiling (since these are laboratories, the duct in the corridor is an exhaust, pulling fresh air in from openings in the curtain wall).

here in New York he also encountered the talents of the one man, in all probability, who could make it work: Willis Carrier. The Conduit Weathermaster system installed at UN was regarded, by Carrier himself, as the crown of his career; neat and sophisticated in itself, it also had to handle unprecedented environmental loads.

As an anonymous commentator cited by *Architectural Forum* put the matter

> . . . air conditioning and venetian blinds are pitted against the powerful sun . . .[14]

[14] Architectural Forum, November 1950, p 108.

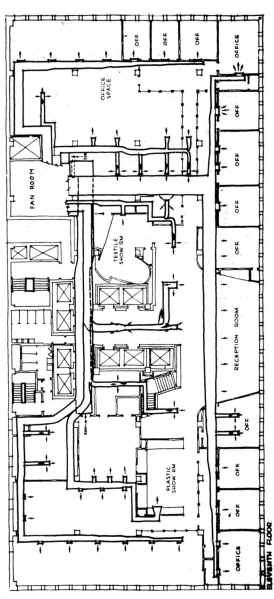

Universal Pictures Building, New York, 1947, by
Kahn and Jacobs; above: exterior; right: overhead
air-conditioning duct plan for a standard floor.

The glazed walls, to Le Corbusier's well-documented wrath, were not protected by any *brise-soleil*, and they faced almost due east and west, with the narrow blank walls to north and south. Defenders of the design have made much of

> . . . the little-appreciated fact that Manhattan Island does not lie due north and south . . . the west wall faces more nearly north west than west, and receives much less sun heat than might be expected.[15]

Not all that much less, however; the published calculations were for 2,400 tons installed cooling capacity for a due-west facing orientation, and only a hundred tons less for the actual orientation. The strongest justifications for the UN's orientation must ultimately be those of convenience in site-planning, and considerations of aspect and view, which should be more than sufficient justification.

Carrier, and Wallace K. Harrison as executant architect, addressed themselves to the environmental consequences of this orientation with the aid of an air-conditioning system that was a true son of PSFS, but far more complex. There is not one intermediate floor of services and plant, but three—at the sixth, sixteenth and twenty-eighth levels, each distributing conditioned air upwards and downwards to the intervening floors—plus a final plant floor at the top of the block serving the floors immediately below, and another in the third basement, to serve the entrance areas and council chambers. These mechanical floors are acknowledged on the exterior of the block by variations in the glazing pattern, but its bland external form makes no other concession to the amount of plant that has to be installed.

Internally, the only acknowledgements are the grilles, diffusers, etc., through which the conditioned air enters the rooms, supplied by a double duct system. Air at low velocity is distributed downwards through diffusers adjoining fluorescent lighting troffers to the internal rooms below the floor/ceiling complex, and at high velocity, upwards through weathermasters round the perimeter of the external rooms of the floor above. Of course, the depth of the floor-slab/suspended-ceiling complex required to accommodate

[15] ibid.

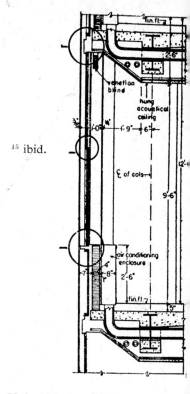

United Nations Headquarters, New York, 1950 (completion of office slab), executive architect Wallace Harrison; above, typical section through wall of office slab; opposite: general view; below: standard outlets in office ceiling.

this ducting is considerable, and would have to show, somehow, if it were to reach the outer glass skin of the block. Fortunately, there is enough floor cantilevered beyond the structural columns—thirty inches, or so—to allow the outer part of the ceiling beyond the column line to be angled upwards at forty-degrees to meet the edge of the floor slab at the skin, and show no more than a thin line—while still leaving just enough volume for the ducts to the Weathermasters to make their necessary upward elbow turn.

This may be ingenious, but it is hardly an elegant solution. It did not occur to Harrison's team to do what Gordon Bunshaft of Skidmore, Owings and Merrill did at Lever House—to double his firewall. The operation of the New York fire code made it necessary to back up a glass skin on the outside of a building with a dwarf wall (usually of cinder-blocks) on the edge of each floor slab to increase resistance of fire spread from floor to floor. In the UN building, this wall is masked externally by a spandrel of obscured glass, on the outside, and neatly accepts the Weathermaster installation on the inside. Lever House accepts a similar basic system for its fire-wall and air-conditioning outlets, but doubles the fire-wall with a similar wall hung below the edge of the slab.

An inch or so shorter than the upstand wall above it, this hung wall provides an excellent mask for the end of a suspended ceiling space amply deep for a full array of services; and the binary nature of this upstand-downstand solution is acknowledged with due honesty on the exterior by a glazing bar which horizontally divides the green glass spandrel into upper and lower panes. Both technically and intellectually it is a more accomplished solution than the UN's, but was not significantly repeated in the later works of the SOM office. Nearly all use a downstand ceiling whose fascia can be seen coming down behind the upper part of the windows, leaving a narrow slot—between glass and fascia—which is sometimes used as a species of superior curtain track, to accept the upper ends of the vertical-slat venetian blinds which are normally needed for sun-control.

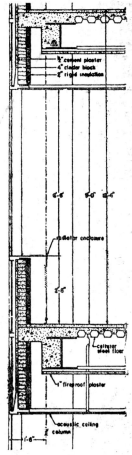

Lever House, New York, 1951, by Skidmore Owings and Merrill; typical curtain-wall section.

Lever House: exterior of office-slab.

Even in this situation, however, the fact that the suspended-ceiling has a fascia to close off the end reveals that the intention is still concealment. By permitting lighting and air-handling to manifest themselves only as rectangular inlets flush with the smooth modular surfaces of the ceilings, by treating Weather-master-type units as mere incidents in the flush surfaces of window sills, the mechanisms of environmental control were hidden from interior view. By housing mechanical plant in standard floors that do little to break the pattern of the exterior glass walls, or by verti-cal prolongations of those glass surfaces above the highest habitable floor levels, these mechanisms were hidden from external view as well. In the full flower of this aesthetic, as represented by the Continental Center in Chicago, by C. F. Murphy and Associates—to take an extremely handsome instance—the presence of mechani-cal plant is acknowledged only by a change from glazing to louvering at the top of the façades. All in all, the aim was to present a smooth rectangular envelope, mechanistic in stylistic preten-sions, but not mechanical in its expressed content. Even when the structure migrated outside the envelope, as in SOM's Inland Steel building, the services did not. And when the air ducts did finally go outside the envelope, as late as 1964 in the same office's Equitable Building in Chicago, they are not seen as such and are boxed into the apparent external structure.

*   *   *

Finally, this chapter on concealed environmental power seems the proper place to acknowledge one of the masterpieces of smooth apparent simplicity and concealment of the 1950 vintage: Philip Johnson's own house in New Canaan. Johnson himself would probably insist that the design is not very original, deriving from Mies van der Rohe's earliest projects for the Farnsworth house and a dozen or more historical precedents that he has listed in print.[16] And he has also said that it is 'not a controlled environment'

[16] *Architectural Review*, Septem-ber 1950, pp 152–159.

Continental Center, Chicago, Ill., 1960, by C. F. Murphy and Associates; exterior showing louvred upper part of façades to conceal servicing plant.

(apparently because it has no air-conditioning). Whatever his personal opinions, the result is discovered by inspection and habitation to be an unique example of environmental management in an extended sense that resumes several themes discussed earlier in this book.

For a start, what Frank Lloyd Wright (who was expectedly rude about the design) would call 'modern opportunities' in servicing have been exploited not merely to spread or 'articulate' the plan, but to split into two distinct units; one the almost entirely solid-walled guest wing which, whatever its internal ingenuities, does not concern the present argument any further,

Architect's own house, New Canaan, Conn., 1950, by Philip Johnson; the glass house by day.

the other the totally-glazed living pavilion—a realisation of Scheerbart's detached veranda, if not by intention—more obviously related to the Farnsworth project, which directly concerns the present argument. Simpler than the Farnsworth design, it presents itself to the observer as an undifferentiated rectangular enclosure of glass, detailed only to the extent of four recessed steel corners, and a full-height door in the centre of each of its four sides.

The glazing is not doubled, so—from the point of view of heat, light, vision and acoustics this is the lightweight wall *in extremis*. A brick drum, suggesting a mechanical core, rises from the floor and passes through the flat roof-slab; it contains a working bathroom, and, on the side toward the main living area, a fire-place which also works—but at the mainly ceremonial level of most of Wright's spectacular hearths. It makes a psychologically satisfying display of combustion, and radiates heat over a limited area. Yet the entire floor plan, even to the most remote glazed corners, is thermally habitable even when snow lies on the ground and against the glass. Heating is, in fact, provided by electrical elements in both the floor slab and the roof slab, and since this requires no visible outlets or registers, the house, under normal conditions when the fire is unlit, provides an ultimate example of invisible heating services.

But in the height of summer it remains equally habitable, and this is the more baffling at first sight because of the lack of any visible sun-controls beyond some internal curtaining. The cooling and sunshading provisions are, however, 'concealed' in full view in the surrounding landscape. The glass house stands on a (partly artificial) bluff projecting from the fall of quite a steep slope that descends from the road at the top of the site to the pool at the bottom of it. The bluff looks west, through a bank of well grown trees rooted at a lower level, and these trees give adequate shade, when in leaf, to the thermally critical south and west walls. Furthermore, the slope and its trees seem to encourage a mildly breezy

local micro-climate even when there is no general wind, so that the opening of two or more of the doors will provide any necessary cross-draught. The same trees, floodlit, also provide a spectacular nocturnal environment even in winter, and the isolation of the house from the public road guarantees visual privacy.

In practice, only two sets of conditions seem to reveal any serious shortcomings in its environmental performance. One is when fine summer weather brings determined architecture-lovers down the drive without a by-your-leave, to interrupt Mr Johnson's privacy. The other is when a very prolonged Indian summer brings low hot sunlight into the house through already leafless trees; at such times the internal blinds are not always adequate to the heat-load of the early afternoon. These few days, however, hardly seem too high a price for a house in the country, to recall Wright's proposition, that is the delightful thing that imagination would have it. Forty years after the Robie house,

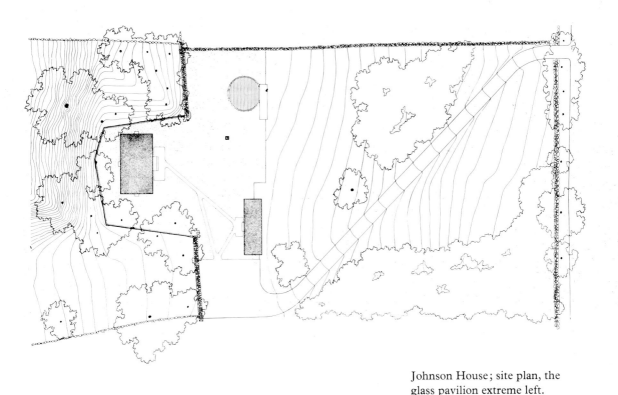

Johnson House; site plan, the glass pavilion extreme left.

Philip Johnson produced (only once, it seems) a masterly re-mixture of mechanical and architectural environmental controls that was as subtle and successful as Wright's. Admittedly, he had innumerable advantages both in budget and site, that are denied to most other architects, but one must still wish that those other architects would more often seize hold of the advantages they do possess with the same imagination and practical craft, and extract more environmental profit from their briefs, budgets and sites.

233

11. Exposed power

The achievement of invisibly serviced glass enclosures clearly satisfied one of the leading aesthetic ambitions of modern architecture, but in doing so it flouted one of its most basic moral imperatives, that of the honest expression of function, and a real conflict of intentions can be felt in the buildings and architectural discourse of the early 1950's. The tradition that had demanded that an electric lamp bulb be manifestly seen as an electric lamp had not died, indeed it was revived by Alison and Peter Smithson who made a manifesto-type point of a naked bulb in an office interior which they designed in 1952 for an engineer friend. So one need hardly be surprised at the fact that later work in one of the first building complexes in which effective concealment of services was first achieved—the United Nations headquarters—should promptly reverse the trend.

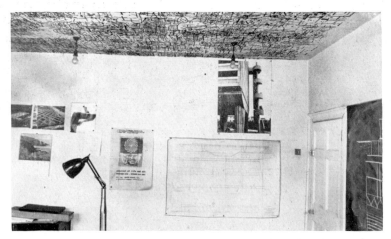

Office interior, London, 1952, by Alison and Peter Smithson; general view showing revival of naked lamp-bulb.

In the specialised council chambers in the podium of the Secretariat tower, overhead services were generally left exposed, picked out in emphatic colours, and—at the most—no more than lightly veiled by a kind of residual skeleton of a suspended ceiling. These interiors were not normally the work of the office of the executant architect, but of architects nominated by the donor-countries which paid for the interiors. Nevertheless, the message did not

United Nations Building, Chamber of the Trusteeship Council, 1952, by Finn Juhl; exposed diffusers and light fittings in ceiling.

United Nations Building,
Foyer of the General Assembly,
1957, executive architect Wallace
K. Harrison; exposed ducting in
ceiling.

go unheeded, and in the giant foyer of the General Assembly building, completed in 1957 (though not as Le Corbusier had intended it), Harrison allowed the massive and complex duct-work of the air-handling system to be fully seen all over the ceiling. It was neither emphasised, nor concealed, but simply allowed to be seen.

Such a solution was almost certainly too casual intellectually to appeal to, say, Mies van der Rohe, who has never proposed anything like it, and too informal aesthetically to appeal to such of Mies's followers as the active designing partners in Skidmore, Owings and Merrill. Historically, the aesthetic blockage is probably the more crucial of the two—the architecture of manifest environmental services had to wait upon a change of aesthetic preferences quite as much as upon the growing difficulties of finding somewhere to hide the ducting and mechanical plant, as both became bulkier and more complicated. And in this change of preferences the influence of Le Corbusier was, as usual, the most important detectable factor. Not that he made any direct contributions to the architecture of mechanical services that are worth mentioning (with one conspicuous exception to be mentioned below) but his movement away from smooth anonymous surfaces, and his growing preference for bulky, plastic and non-rectangular forms, both helped to create a climate of taste in which far greater freedom of architectural expression became possible, a situation in which mechanical services often provided the impetus or excuse for expressive formal experiment.

His heroic and sculptural foul-air stacks on the roof of the *Unité* at Marseilles must be acknowledged as historically important if only as the first explicit sign for almost twenty years in his work that mechanical services are an expressible function of a building. But it must be admitted, also, that their direct influence on the general manner of designing buildings was practically zero—their only well known descendant is the smoke-stack from the heating plant of the 'little *Unités*' in the London County Council's housing

Unite d'Habitation, Marseilles, France, 1952, by Le Corbusier; foul air stack on roof.

development at Roehampton, and even this simplifies the form. On the other hand, a search through competition entries and student thesis projects of the middle fifties would certainly show many more examples, and even more examples still of influence from his design for Notre Dame du Haut at Ronchamp. Although this building has nothing ostensibly to do with the architecture of services, its silhouette—with informally grouped towers rising above a massively overhanging roof—was one that could be obviously and directly adapted to situations in which vertical and horizontal services had to be accommodated—an example is the Queen Elizabeth Hall in London, which will be discussed at the end of this chapter. But even more than this, it was Ronchamp's total rupture with the tradition of regular rectangular building-forms that was to be most consequential.

This is clear if one examines any quantity of the projects which seem to derive from Ronchamp, and make expressive use of servicing provisions. Often it will be found that though they are picturesque and irregular in plan or silhouette, the detailed forms are still rectangular, sharp-edged and without Ronchamp's free-hand curves. This is the more striking when one reflects that air-handling equipment, whose sheer bulk makes it the most likely candidate for the dubious role of 'technological determinant of form', has an extremely characteristic repertoire of stepped and tapering shapes, not to mention the curved scrolls of fan housings, yet these have hardly ever figured in the repertoire of the 'architecture of services.' Far from being a determinant of form, environmental machinery has tended more and more to become the stimulant or excuse for experiments made possible by the liberating effect of the products of these environmental machines.

In freeing architecture from local climatic constraints, mechanical environmental management techniques have given *carte blanche* for formal experimentation. For instance, the glazed rectangular slab block is in many climates and locations a far more wilful and romantic proposition (from the point of view of locally accepted vernacular traditions of what is, or is not, environmentally tolerable) than some picturesque and irregular format might be. By making almost any kind of building form habitable at almost every point of the world's surface, environmental services have made every kind of building look faintly exotic, wherever it appears.

It is therefore quite in order, probably, that the first building to be considered in this chapter was designed in Milan for a site at Merlo in Argentina, and would look equally at home in either place. It forms one of a (purely chronological) cluster of designs which may be regarded as the heralds of the present phase of the architecture of services. The other two are Franco Albini's *Rinascente* store in Rome, and Louis Kahn's laboratories in Philadelphia. All three were in the process of design towards the end of the fifties, but Zanuso's factory for Olivetti-Argentina had reached its definitive form by 1959, and thus deserves to be considered first.

The general conception of this plant was entirely conventional: continuous single-storey shedding. However, it seems to be in the nature of single-storey structures to facilitate experiments in environmental control—the Royal Victoria Hospital at one historical extreme, serviced throughout from below, or SCSD at the other historical extreme, serviced throughout from above. Zanuso's solution is an overhead one: structurally, the work-space is sheltered by an elaborate membrane consisting of alternate wide roof planes set high and narrow ones set low, the difference in height providing continuous monitor lighting. The high roof planes are supported on transverse beams carried on tubular girders lying in the lower planes which are, effectively, projecting horizontal fins on either sides of the tubes. The whole is in reinforced concrete, the span from one tubular girder to the next being just over twelve meters; the tubes in their turn being supported at intervals of eighteen metres by columns of a complex 'cross of Lorraine' section, beyond the last of which in any row the tube cantilevers a maximum of three metres.

Olivetti factory, Merlo, Argentina, 1964, by Marco Zanuso; air view of factory.

Olivetti factory: closer view
showing ends of hollow tubular
girders.

The monitors are left unglazed in some areas where they serve
simply as car-ports or shelters, but where total enclosure is required, the monitors are glazed and the perimeter sealed by simple
glazed walling in either direction on the line of the column
centres—which means that there is always some structure cantilevering beyond the glass to shelter it, whether it be the fin on the
side of a girder along a side wall, or the projection of the higher
roof-planes beyond the girder-ends on an end wall. Where such

end-cantilevers occur, the upper planes are extended to join edge to edge, suppressing the monitors, to form a continuous canopy. This serves to keep sun and rain off the end wall and, more importantly, off the environmental mechanisms.

For, where air-conditioning is required, it is provided by exposed units, hung from a steel chassis cantilevering above the girder, and using the girder's hollow interior as a duct-space for air distribution. Outlet slots are provided at intervals in the lower face of the girder, under which most of the piping and conduitry are hung as well, and air return to the conditioner is provided by an exposed duct emerging through the central web of the column and turning upwards into the underside of the conditioner units. This classic 'clip-on' solution, in which the environmental power is applied to the building almost in the manner in which propulsive power is applied to a boat by outboard motors, not only makes the conditioning units immediately visible and accessible for servicing, but also seems to satisfy a deep intellectual and moral need: the need to be able to see the difference between the structure, which is supposed to be permanent, and the services, which are hoped to be transient, and to see that difference made expressive. The building is serviced, and manifestly seen to be serviced; the fact of servicing is seen to be within the architect's control, even if what is seen is not, in detail, entirely of the architect's design.

Neither Kahn's, nor Albini's designs possess this frank and gratifying clarity. Zanuso enjoyed the traditional advantage accruing to anyone who designs a factory—that such buildings are not felt, even now, to be serious representational architecture, and the cultural restraints are therefore that much the less severe. In Albini's case the cultural restraints would have been—to a Milanese of his generation—almost crushing; he was designing a building for a conspicuous site in the history-sodden *ambiente* of Rome, at a time when the historical nerve of most Italian architects had failed almost completely (these were the years of Neoliberty nostalgia).

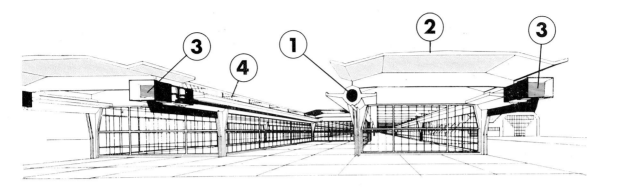

Olivetti factory; above: perspective diagram of structure; left: air-conditioning unit (redrawn from *Casabella*).

1. Exposed end of hollow concrete beam
2. Main roof structure
3. Air conditioner unit attached to end of hollow-beam duct
4. Monitor lights in roof

The basic proposition of the *Rinascente* store is simple enough —a windowless multi-storey box of selling-space, standing on virtually an island site, with almost its entire exterior surface available for the development of services. The realisation of this simple proposition became increasingly complex as the design proceeded from the first to the final version, however. The first version with its top-heavy silhouette and roof-top parking space, boldly exposed steel frame and conspicuous exterior staircase, suggests an inspiration from early industrial plant or, according to Fello Atkinson, Futurist memories.[1] The tempering of that original inspiration to meet the supposed demands of the *ambiente preesistente*, shrank that exuberance into a classical silhouette for which there is direct local precedent—a six-storey apartment block close at hand in via Salaria.

So, what Albini finally offered as a 'technological building for an historical setting', turns out to be a nineteenth-century palazzo with a low-pitched roof and its classical detailing reworked into a finnicky and elaborate exposed steel frame. The infill to that frame is of pre-cast concrete units with a variegated exposed red aggregate (and a single white string-course halfway up each storey-height). These infill panels are not flat, but are given a broad corrugation (some of the folds are three feet wide) by projections which house services such as air-trunking and pipe-runs, descending from the plant-rooms under the roof to the floors below. Typically, an air-conditioning duct will descend to the level of the floor-slab above the storey it is intended to serve, and there, passing behind the projecting steel cornice, is turned back inside the building, passes above the pair of girders which form the 'frieze' below the cornice, and then turns down to serve a ring duct running around the perimeter of the plan, whence it distributes air downwards through registers in the ceiling of an interior scheme which is not—worse luck—of Albini's designing.

Since the duct has been turned back inside the building, it needs no accommodation in the pre-cast skin below the cornice, and the

[1] *Architectural Review*, October 1962, p 270.

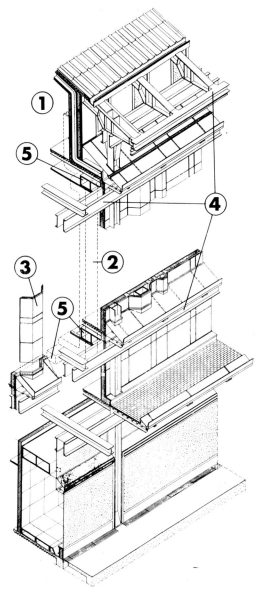

La Rinascente Store, Rome, 1961, by Albini and
Helg; above: detail of corner; right: cut-away showing distribution of ducts and services in outer wall
(redrawn from *Casabella*).

1. Plant room in roof
2. Vertical distribution duct
3. Pre-cast cladding
4. External steel framing
5. Distribution duct to sales space

corrugation which housed it is suppressed accordingly. Thus the wall of the uppermost storey is the most extensively corrugated, while that of the first floor is almost innocent of such projections. There is an irony in this, for many visionary projects of the immediately preceding years had made a downwardly-attenuated and externally-exposed system of duct-work the excuse for a dramatically top-heavy expressionist silhouette, whereas *La Rinascente*, almost the only building with this type of servicing to be erected, constrains the servicing system within so rigid, yet timid, a classicising format that it goes almost unnoticed.

This is undoubtedly a pity, for *La Rinascente* has not had the amount and type of discussion it deserves. Conceptually, it is important as a very clear demonstration of a building skin performing environmentally in a double role: passively as a static barrier to the entry of external climate or the loss of internal climate; actively as the distributor of conditioned air and environmental power. But such subtleties, however visible to the attentive viewer confronting the building where it stands, do not register in the graphic representations on which architectural discourse is too often based. The corrugations of the external wall do not register on the ground plan because the services do not descend to that level, and they are therefore missing from the only document to which some architects will ever give serious attention, and they are too unemphatic to make much showing on the average magazine photograph or colour-slide, which tends to emphasise instead the historically craven silhouette.

Not so the Richards Memorial laboratories: there, Louis Kahn's apparent provisions for environmental services give an immediately striking profile to both plan and elevation, and have been equally immediately understood and admired. No building in recent years has presented such an air of novelty on the basis of planning methods that were so old—it is worth noting that it was for this, and not for *La Rinascente*, that *l'Architettura* coined the term *Arcaismo Technologico*.[2] Kahn's design offered an immediately

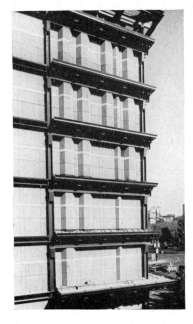

La Rinascente; above: detail of wall; right: general view.

[2] *l'Architettura*, October 1960, p 410.

Richards Memorial Laboratories, Philadelphia, Pa., 1961 by Louis Kahn; laboratory towers side.

comprehensible solution (i.e., one already familiar in some way) to an increasingly pressing problem—the proper servicing of scientific work-spaces. However unique the site whose constraints— between the Medical School, Zoology Buildings and Botanical Gardens of the University of Pennsylvania—are supposed to have dictated the choice of clustered towers, this solution was taken to be universal and general, and imitation was so instant and so widespread that Colin St John Wilson had to enquire (*Perspecta VII*) 'Will "servant spaces" be the next form of decoration?' And so deep and irrational is the reverence in which the design is held, that some of its admirers will express themselves pleased at the deficiencies of the design and claim the fact that 'it doesn't work' as the ultimate proof of its architectural worth.

Doubtless these exaggerated responses stem from an uneasy relationship between the technology of US architecture and the cultural contexts in which it is customarily discussed, whether in criticism or teaching. The concept of architecture as a perennial art often consorts ill with the transient-seeming facts of piping and ducting and wiring. Hence Le Corbusier's complaint at the luck of Ledoux in not having to cope with piping; hence too, presumably, Kahn's own despairing account of the design of the laboratories:

> I do not like ducts, I do not like pipes. I hate them really thoroughly, but because I hate them so thoroughly, I feel that they have to be given their place. If I just hated them and took no care, I think that they would invade the building and completely destroy it.[3]

[3] quoted in *World Architecture I*, London, 1964, p 35.

While the non-architect must wonder how it can be that a man so thoroughly out of sympathy with more than half the capital investment in a building of this kind should be entrusted with its design, the consequences of his lack of sympathy are plain to see. To stop the pipes and ducts he so hated from destroying his building, Kahn gave them their place outside those volumes which he appears to have considered to be 'the Building'. The laboratory spaces ('the building', presumably) occupy variously subdivided floors of short (7-storey) towers of square plan, forty-five feet to a side. The three lab-towers which formed the first instalment of this phased design are grouped around a somewhat taller, and mostly unglazed, core-tower of ancillary spaces.

Air for ventilation enters four upcast ducts which project from the rear of the core-tower (as it were the Larkin Building situation upside down), and is drawn up to a plant room at the top. Thence, conditioned air is blown down two massive distribution shafts

Richards Laboratories: upcast ducts for air intake.

buried invisibly in the core tower, and into the work spaces, and vitiated air, together with other wastes, is extracted through the blind brick turrets on the centres of the exposed sides of the lab-towers. It will now be understood that not all the mechanical services are in these external turrets, as is commonly believed, since some are in the core-tower. Furthermore, not all the visible 'servant space' turrets on the exterior contain services; at least one on each lab-tower contains an escape staircase. In early

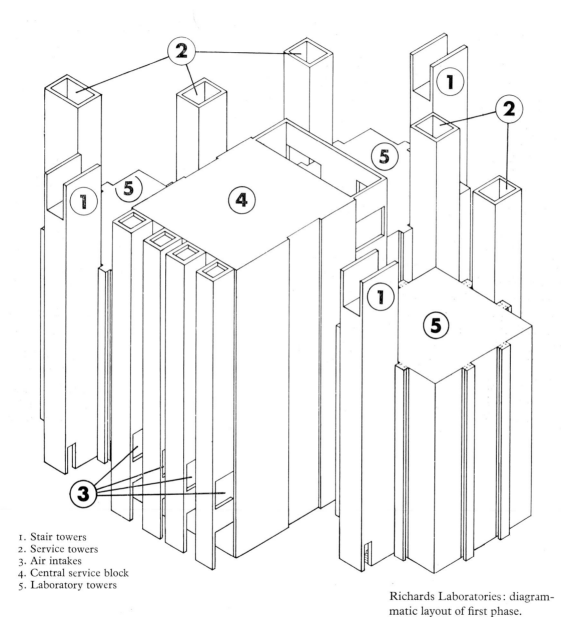

1. Stair towers
2. Service towers
3. Air intakes
4. Central service block
5. Laboratory towers

Richards Laboratories: diagrammatic layout of first phase.

versions of the design, the escape stairs were in cylindrical towers that contrasted strongly in form with the ribbed, square service towers, but in the final form, stairs and services alike were housed in square brick turrets of almost identical form.

As the building strikes the eye, then, it consists of glazed served towers surrounded by blind servant towers, a *parti* of almost Beaux-Arts simplicity, and crudity. Effectively, what Kahn has done is to provide the laboratories with monumental cupboards in which all the services he hates can be forgotten because outside the plan of 'the building'. Where these pipes and ducts cannot be kept out, that is, where they enter the workspace to distribute their services, he encountered difficulties that could not be tidied up in the same way, and which he did not want tidied up in the ordinary commonsense way. He proposed to thread his services through the open interstices of the concrete truss system which carried the floors, without a suspended ceiling beneath, which suggests that he was still honestly trying to grapple with his problem. But such a solution did not recommend itself to the users of the laboratories. Not only did their normal aesthetic expectations include a flush ceiling, but their demands for acoustic privacy and absolute cleanliness could not be met by open and undustable concrete truss-work.

These, however, were purely local tribulations, however painful to both parties; the world-wide consequences of the design had only to do with the exterior, and the way in which this made the pressing problem of services capable of being discussed in the traditional terminology of massing and plan. Historically, the functional and environmental qualities of the finished building, and its conceptual misfires (Arthur Drexler: 'Of course, the labs should have had solid brick walls, and the service towers should have been light glass construction for accessibility and change')— historically these will count for nothing against the effect of the whole complex in offering an instant solution to a problem, and in bringing that problem within the terms of customary architectural

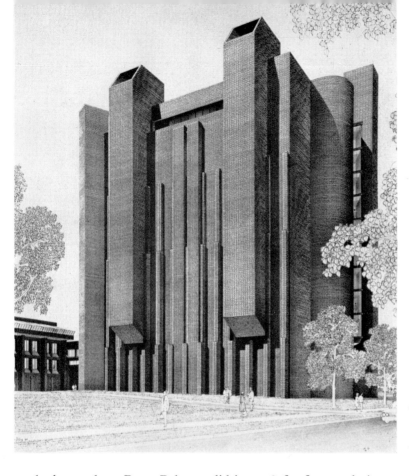

Agronomy Laboratories, Cornell University, N.Y., project of 1965, by Ulrich Frantzen; exterior.

method—much as Peter Behrens did in 1908 for factory design, with his *Turbinenfabrik.*

Too much of the method derived from the Philadelphia laboratories is, as Colin St John Wilson feared, purely decorative, even empty in a literal sense—often the projecting 'servant spaces' of published projects and competition entries prove to be pure pseudopodia of the external wall, containing not even staircases. Very few architects have even felt compelled to go back and work through Kahn's design again to see if, in fact, it could have been made to work as a laboratory building. Ulrich Frantzen's project

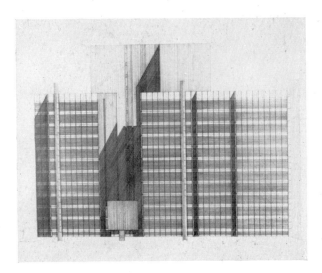

Above right and left: Pharmaceutical plant (project), Debreczen, Hungary, 1962, by Gulyas and Szendroi; two views.

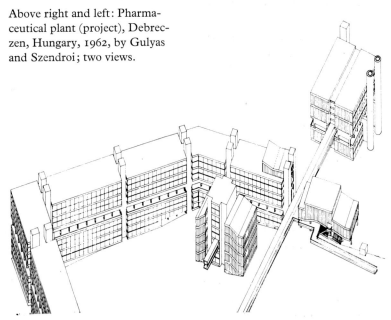

Left: Sheffield University extensions (competition entry), 1953, by Alison and Peter Smithson; detail showing service towers.

for an Agronomy laboratory tower at Cornell University is not merely the only published version that is worth a second look, it is virtually the only published version. The apparently similar solution to apparently similar problems in the design for a pharmaceutical plant at Debreczen in Hungary, by Zoltan Gulyas and Jeno Szendroi, owes a great deal to Kahn visually, but has only staircases in its visible servant towers; the riser shafts for airducts, though nominally external to the floor plan are so buried in the convolutions of the external envelope that they are not seen.

And there is yet another historical consequence of this paradoxical building. Simply by being built it legitimised, so to speak, a number of ideas about exposed services that had been floating about in that underground world of student projects and forgotten competition entries to which reference has already been made. Some of these ideas are relatively ancient by the standards of the present chapter. Thus, as early as 1953, something of the architecture of the Richards Laboratories had been anticipated in the competition project for Sheffield University by Alison and Peter Smithson. Indeed this project went further; not only does it propose at least one glazed tower that is flanked by servant towers, but it extends the concept of externalised services horizontally, moving piping and persons from building to building in double-decked ductways, piping above, persons below. This scheme had been published as early as 1956, but it is unlikely to have had any influence on Kahn, and by the time he was working on the Philadelphia labs, the English underground had moved on, to a position exemplified by Michael Webb's even better-publicised student project for a 'Furniture Industry Headquarters Building', seen as a series of free-form capsules suspended in an exposed frame and connected by external tubes which may or may not contain people, or ducts, or other services.

At the time this chapter is being written, the English underground with its futuristic and mechanistic dreams is a matter of international comment, because of the reputation of the magazine

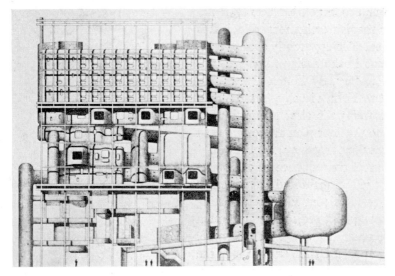

Furniture Industry Headquarters (student project), 1960, by Mike Webb; elevation.

Archigram. The group which produces the magazine and its associated manifestations includes Michael Webb, mentioned above, and also the effective designers of the next building to be discussed, albeit the design antedates their membership of *Archigram.* Indeed the Queen Elizabeth Hall complex on the South Bank in London would have looked very much as it does, in all probability, had the *Archigram* connection never come to exist. It was designed in the early 1960's by a design team in the Special Works Division of the LCC Architect's Department, and municipal protocol requires that credit for its design be distributed down the hierarchy from Hubert Bennett, Architect to the Council, to Norman Engelback, who was 'Group leader for the project.' It is widely known, however, that the 'real designers' (a conveniently loose term) had been Ron Herron and Warren Chalk with, later, Dennis Crompton—all three to become among *Archigram*'s most active visionaries.

When the concert-hall parts of the complex were completed, in

1967, they were naturally enough scrutinised fairly closely for signs of *Archigram*'s 'Plug-in' aesthetic, which by then was well known, whereas attention, when the model had been published some years earlier, had concentrated on allegedly Corbusian elements, such as the use of exposed concrete on the exterior. In truth, one could say that the Corbusian and Plug-in elements are manifest in one and the same thing, the silhouette the buildings derive from the external disposition of the main service ducts. That silhouette is 'romantic' in the conventional everyday sense of the term, and the whole exterior presentation of the design conspicuously lacks either the diagrammatic clarity of Kahn's laboratories, or the pre-figured compactness of *La Rinascente*.

The cluster consists of four main parts—two performing chambers, a foyer serving both, and a plant room serving all three. Visually, it is extremely difficult to distinguish these parts at first; they have been mixed and overlaid in a manner which suggests both a picturesque intention and a much more relaxed attitude to piping and ducting (and services generally) than either Kahn's or Albini's. Given this more comfortable technological stance (and the support of intelligent engineers within the parent organisation) the architects were able to propose an architectural solution that would do two things: satisfy the exacting environmental requirements of its internal functions, and make architecture out of the provisions needed for their satisfaction. Thus, while the free-form foyer has a largely glazed perimeter, the two performing chambers, which in fact lie neatly at right angles to one another, have totally blind exteriors, in order to extract the maximum sound insulating performance from their solid fifteen-inch concrete walls.

Since they are closed boxes perforated with the minimum number of door openings, the performing-chambers must have totally artificial atmospheric management. This is provided from the plant room which straddles the smaller chamber on an independent column-structure. Air intake grilles and upcast exhausts form conspicuous features of the exterior of the plant-room, and so do the

ducts for the conditioned air—though all one sees in fact from the outside are the concrete casings in which these intakes, ducts, etc., are located. One such casing runs round the upper part of the plant room like a massive cornice, and contains the extract duct from the smaller chamber, its branch-ducts corrugating the external wall almost like that of the *Rinascente* store. A second and larger duct bridges the gap between the two chambers and crosses the upper part of the larger one to the opposite side, where it is

Queen Elizabeth Hall (South Bank Arts Centre), London, 1967, by the Architect's Division, London County Council (later, Greater London Council); general view with plant-room at upper right.

Queen Elizabeth Hall: plenum feed duct, right, and projecting duct-housings on exterior.

divided and wraps around the entire perimeter of the chamber, again in a casing like a cornice.

A third conspicuous distributor shaft is taken from the top of the plant room, bridges the gap between it and the foyer in mid-air, then turns down and enters the top of a sizable duct-housing that runs along most of the upper part of the inner wall of the foyer, in order to distribute air for the Plenum system that ventilates this part of the cluster. Although most of the return ducting from the

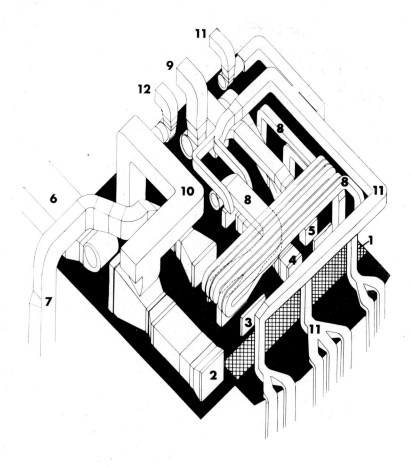

Queen Elizabeth Hall: cut-away
of air-conditioning plant room.

1. Fresh air chamber
2. Separate fresh air supply for
   main auditorium QEH
3. Air supply for foyer Plenum
   system
4. Air supply for Purcell Room
5. Air supply for plant room
6. Main splitter duct to QEH
7. Down to foyer
8. Purcell room supply across
   plant room and down to ceiling
   below
9. QEH extract
10. QEH re-circulation
11. Purcell room extract
12. Plant room extract

air extracts under the seats of the two performance-chambers is
naturally buried in the bulk of the building and would thus be
difficult to acknowledge externally in the same way, there can be
few public buildings in which the main primary air-distribution
is made so rhetorically manifest.

The word 'rhetorically' is used advisedly, since the making
manifest has a large element of symbolism in it. Although the

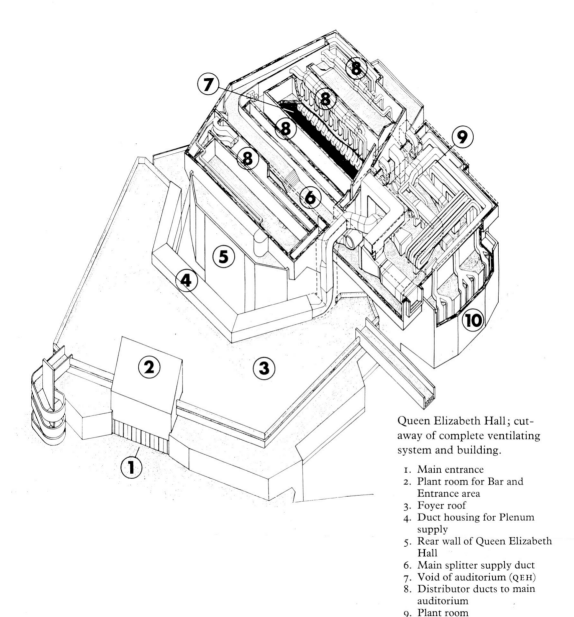

Queen Elizabeth Hall; cut-
away of complete ventilating
system and building.

1. Main entrance
2. Plant room for Bar and
   Entrance area
3. Foyer roof
4. Duct housing for Plenum
   supply
5. Rear wall of Queen Elizabeth
   Hall
6. Main splitter supply duct
7. Void of auditorium (QEH)
8. Distributor ducts to main
   auditorium
9. Plant room
10. Rear wall of Purcell Room

Queen Elizabeth Hall; above and opposite: two views of interior of main chamber, showing air inlet ducts in ceiling.

externally visible casings (normally distinguished by their elegant *in situ beton brut* from the inhabited spaces, which are clad in pre-cast panels) do in fact contain air on the move, it should not normally be assumed that what is seen from outside is necessarily the form of the ducts through which that air is moving. Most usually, the visible concrete work gives only an approximate idea of the true forms of the metal ducting within, which in many cases is quite a loose fit inside, with room for service engineers to crawl past it. Nevertheless, the general external form takes its cue fairly directly from the facts of air-flow within.

Inside the performing-chambers, the larger one for instance, conditioned air enters from diffusers in rows across the ceiling, set in visible concrete panels which correspond to the duct-containing corridors above. The diffusers are in the form of very visible ventilating elbows which turn toward the back of the hall, so that ventilation is manifestly seen to be done, while an ingenious system of roller-shutters enables the diffuser grilles in the wall at the back of the stage to rise and fall with the stage as it is adjusted for differing types of performance. The illumination of the stage area is also by manifest and visible means, consisting of studio-type lamps on battens crossing the whole upper part of the stage. This is not affectation, since the stage can be used as a television studio, but full value, visually, has been extracted from the fact that lighting of this standard has to be provided.

The way in which usages such as these have been described as 'perverse' by English critics—those on the exterior even more than those within—in spite of their fundamental good sense, shows how far the general body of architectural discourse has to go in coming to terms with 'the architecture of services'. This has nothing necessarily to do with whether or not the building 'works'—critics have tended to be confusedly indulgent about the environmental difficulties of its early days, when the very long time taken to warm the massive concrete structure and its exposed surfaces led to difficulties with unmanageable down-draughts, etc., which would not have occurred in an auditorium with less exposed cold concrete about; nor with the usual disputations about acoustics which are stirred up by any new auditorium, though such considerations of acoustic and environmental performance must enter any definitive judgement on the building. What is at stake here is the tendency to offer definitive judgements on the basis solely of visual inspection, coupled with expectations derived from the visual inspection of buildings that set about solving their problems in an entirely different manner. Since most of our experience and expectation at present derives from buildings that do not deploy totally mechanical environments (because such installations are still rare) we are not yet in a position to hand down confident judgements on them. They are the fruit of a revolution in environmental management that is without precedent in the history of architecture, a revolution too recent to have been fully absorbed and understood as yet, and a revolution still turning up unexpected possibilities, as the next chapter will show.

12. A range of methods

The unprecedented history which has been sketched in the previous chapters, can be summed up in two ways: either as the final liberation of architecture from the ballast of structure, or its total subservience to the goads of mechanical service. Both interpretations of the situation are current, largely because of the infantile fallacy that architecture is necessarily divisible into function and form, and that the mechanical and cultural parts of the art are in essential opposition. The division also typifies the split between the generations of architects—now and right back through the twentieth century, the sign that an architect was achieving 'maturity' and success was that he had tacitly, or noisily, abandoned the attempt to extract symbolic values and cultural performance from the application of advanced technology—Le Corbusier abandoned the attempt around 1933, but the *Archigram* connection are still trying. Indeed, they were threatening at the end of *Archigram 7* that

. . . there may be no buildings at all in *Archigram 8*.

Such willingness to abandon the reassurances and psychological supports of monumental structure are rare—a notable exception to the general rule, for instance, is Buckminster Fuller who has always expressed hostility to sheer mass in what he would call 'the shelter industry.' Although Fuller is now one of that industry's senior citizens, his birth in 1895 meant that he entered the world almost exactly half-way between the introduction of domestic electric light and that of industrial air-conditioning. An enthusiastic child of his time, he has always been at home in the world that these two epochal innovations have transformed.

The profession of architecture (of which Fuller is only an

honorary member, on the basis of 'Join him, we can't beat him') as a continuing body of human activity, is not a child of that time. Its traditions, as a conscious intellectual discipline, go back to the Italian renaissance; and as a practical craft they go back almost to the dawn of human culture. Conditioned to admire structures that have stood two-thousand years or more, and required by social habit to design in terms of centuries, that profession has been ponderously slow to change its mind or re-formulate its attitudes; it has tended to believe itself in the throes of major revolutions when confronted with technical innovations that other crafts and disciplines have taken in their strides. For a demonstration of this difference, one can profitably revive Paul Valery's contrast between Eupalinos, the architect (from his Platonic dialogue of the same name) and Tridon, the shipwright. The former was preoccupied with the right method of doing the allotted tasks, and deploying the accepted methods, of his calling, and seemed to find a philosophical problem in every practical decision. Tridon, on the other hand, applied every technology that came conveniently to hand, whether or not it was part of the shipbuilding tradition, and treated the sayings of philosophers as further instruction on the direct solution of practical problems.

Applying a similar comparison simply as an historical test to what has happened to ship-building and architecture, since the emergence of modern technology, we find: from the *Pyroscaphe* of 1783 to the developed hovercraft of today, a process of continuous innovation and broadening choice of methods, compared with which the architecture taught in schools has reached a condition analogous to that of a sophisticated yacht with glass-fibre hull and aluminium mast and other improvements to structural materials—and an outboard motor to be used (under conditions of great embarrassment) in emergencies. If no other possibilities in navigation were possible, we might well prize such beautiful boats as the consummation of art and technology. But they are not *the* consummation, merely one of a range of consummations, for there are stepped hydroplanes as beautiful in shape, hydrofoils as compelling in motion, cushion craft handier and safer in shoal waters, and a whole range of sub-aqua vehicles that can go where a yacht cannot, and would be helpless if it could.

Now it might be argued that conditions on land are less extreme than those at sea, and that there is not the same sheer compulsion of physical survival to drive on the architect to the kind of continuous innovation that is manifest in the design of craft that must weather the intemperance of the oceans. Such an argument—and it is of a class of arguments often advanced in defence of architecture—ignores one damaging fact: that outside the culturally protected circle of what is taught in architecture schools and discussed by architectural pundits, there has been for almost a century a tide of innovation in environmental management fully comparable with that in nautical design.

The sweep and magnitude of that tide has dragged architecture with it, willy nilly, as the previous chapters will have shown, but architecture, as a body of skills, has long since lost control of it— architects as an organised profession have been happy to hand over all forms of environmental management, except the structural, to other specialists (electrical, mechanical, engineers; heating and ventilating specialists; consultants on traffic and systems engineering, communications and control) and they have taught young architects to continue this dereliction of manifest duty; most third-year architecture students can calculate a simple concrete structural frame but very few know how to begin calculating a solar heat load.

It is obviously too late in the day to begin blaming architects for the fact that this situation exists, especially since the blame lies also with society at large for not having demanded of them that they be any more than the creators of inefficient environmental sculptures, however handsome. But we have to face the fact that the architect as we know him at present, the purveyor of primarily structural solutions, is only one of a number of competing environmentalists, and that what he has to offer no longer carries the authority of either necessity or unique cultural approval. In an increasing number of situations that were formerly thought soluble only by the erection of a building, workable alternatives are, for a variety of technical reasons, now becoming available. The obvious and most often cited example is that of the drive-in moviehouse, which is no house. With its audience bringing their own environmental packages with them in the form of automobiles, the need for a permanent enclosing structure disappears, the task of the designer (who may or may not be an architect) is to devise a combination of landscaping, traffic engineering, electronics and optics, plus a modicum of weather-protection for the projection equipment.

If this is a special case, it is considerably less specialised than that of the hovercraft, and vastly less so than that currently-obsessive paragon of environmental management, the space capsule. And the fact that it can be dismissed as a special case at all by architects shows how desperately their vision of their function in the world has narrowed in relation to the means available for performing that function. Isaac Ware's *Complete Body of Architecture* covered almost the total technology of environmental management available in 1750; the practical parts of Guadet's *Eléments et Theories*, amounting to about three times the bulk of Ware's work, cover considerably less than half the environmental technology available in 1900, simply because, like Ware, Guadet was discussing almost exclusively the structural part of environmental technology, which was already less than half the kit.

Perhaps the most damaging criticism that can be made against the established guardians of architectural culture is that it is rarely they who bring new aspects of environmental management into the general body of discourse, but outsiders who force them upon their attention. The most spectacular recent example of this has been the way in which it has needed a literary man, operating on the very fringes of currently acceptable 'culture' to propose a term of comparison by which the large scale manipulation of the nocturnal environment can be related to the accepted body of architecture. And if Tom Wolfe's comparison between Las Vegas and the Palace of Versailles[1] shocks architectural opinion, it is less because it was meant to startle, than that it would never have occurred to any architectural critic (including the present author, I must admit) that the two entities were comparable. And the difficulty in conceiving of a comparison between what was created at Versailles for *le Roi Soleil*, and what was created at Las Vegas at the behest of Buggsy Siegel, stems largely from the contrast of the means employed, not the praiseworthiness or otherwise of the intentions behind their creation.

The difference of means is this: at Versailles the enclosure of space by massive structure is paramount, and the idiom thus created sets the cues for the manipulation of space by other means, such as planting and water; whereas at Las Vegas, structure is the least dominant element in the definition of symbolic space. What defines the symbolic places and spaces of Las Vegas—the super-hotels of The Strip, the casino-belt of Fremont Street—is pure environmental power, manifested as coloured light. Whether or not one agrees that the use made of that power is as symbolically apt as the use made of structure at Versailles, the fact remains that the effectiveness with which space is defined is overwhelming, the creation of virtual volumes without apparent structure is endemic, the variety and ingenuity of the lighting techniques is encyclopaedic. And the scale of the operation is as overwhelming as that of the officially admired monuments of nineteenth-century construction, such as the Forth Bridge, or of Baroque planning, such as Versailles or Sistine Rome. And in a view of architectural education that embraced the complete art of environmental management, a visit to Las Vegas would be as mandatory as a visit to the Baths of Caracalla or La Sainte Chapelle.

The point of studying Las Vegas, ultimately, would be to see an example of how far environmental technology can be driven

[1] *The Kandy-kolored Tangerine Flake Streamline Baby*, New York, 1965, pp xvi–xvii. The specific point of comparison was the architectural consistency of Versailles and Las Vegas, but the general tone of Wolfe's observations on Las Vegas all through the book suggest that the comparison may be taken very much further.

beyond the confines of architectural practice by designers who (for worse or better) are not inhibited by the traditions of architectonic culture, training and taste. But this is not to say that architects themselves have not made forays beyond the common confines of their calling, especially in the field of exhibition design. Even so, their use of light has remained pretty timid by the standards set in practice by Las Vegas, or in theory by Paul Scheerbart—the change from forms assembled in light to light assembled in forms is still too big for most of them. But in other aspects of the application of environmental power, one can point at a certain number of exhibition environments where the imagination of the architect has matched the promise of the technology.

A fairly well-known, though insufficiently studied example, is the demountable pavilion designed for the US Atomic Energy Commission by Victor Lundy, architect, with Walter Bird of the Bird-air Corporation, specialists in inflatable structures of this type. The project is notable, firstly, for its early date[2]—while inflatable structures have acquired a considerable vogue in the student underground since the mid-sixties (partly for their formal qualities and partly because of the prestige of their main proponent, Frei Otto) the AEC pavilion was in public use at Rio de Janeiro as early as 1959, and has seen service in various parts of the world in the ensuing near-decade.

Besides its durability, it is notable among inflatable structures for its size, complexity and open form on plan. Whereas most air-supported structures tend to be simple domes, or elongations of domical forms that still retain a closed figure in plan, the AEC pavilion is better described as an open-ended vault, or half-tube, deformed to produce two approximately hemisperical spaces joined by a central neck, and entered by means of arched porches, about the same diameter as the neck, at either end. Internally, there is a smaller inflatable dome to house a model atomic reactor, and sundry rigid, non-inflatable partitions, projection screens, and so forth. The precise distribution of credit for the design between

[2] the date is early in terms of the practical technology of air-supported structures, not of their absolute invention. British patriots are apt to make much of Dr Lanchester's patent for an air supported structure, filed in 1917, but an interval of almost forty years elapsed before workable structures of this kind could be made and marketed—in the first instance by Walter Bird, whose company was founded in 1956, some ten years after his first successful experimental models.

Opposite : Fremont Street, Las Vegas, Nevada; an environment defined by artificial light.

Lundy, Bird and the consulting engineers, is not easy to fix, but the result remains virtually the only air structure to date with any pretensions to architectural sophistication.

It is also technically sophisticated and complex, employing—as all but the most elementary air-domes must do—a number of techniques to ensure structural stability. Thus, the open porches, being outboard of the revolving doors which act as air-locks to maintain the general support pressure within, must be supported by some other means, and are in fact pure inflated structures, air balloons kept rigid by internal pressure, rather than supported on a cushion of air. The main volume—some 230 feet long, fifty feet to

United States Atomic Energy Commission Portable Theatre, 1959, by Victor Lundy and Walter Bird; exterior.

its highest point and over 100 feet wide at the widest—is a true air-supported structure, however, between the airlocks which seal in its supporting pressure of 49 mm. above atmospheric. But if this is a true air-supported structure, it is not a simple one, in that it partakes of some of the nature of an inflated structure as well—it has a double skin, and some pressure is maintained between the two skins. This offers protection against sudden collapse due to accident (a risk with all single-skin air-supported structures) or vandalism (a risk to all US buildings in underdeveloped countries).

The true point and justification of this building in the context of the present study is that it makes unmistakable architecture out of the exploitation of a new technology, not in the sense that it is made of a new material, or that its components were fabricated in a new way (the two propositions that appear to exhaust the concept of technological innovation in most architectural discussions) but because it consists of a weatherproof membrane incapable of

Atomic Energy Commission Theatre; revolving doors at entrance.

supporting itself, but not—like the membrane of a tent—supported on a discrete rigid frame or tensile structure. The cushion of air which does support it exists only through the constant ministrations of environmental power, the operation of a small air-pumping device—in this case the air-conditioning plant. It thus presents us with a total reversal of traditional roles in architecture and environmental management. Instead of a rigid built volume to which power must be applied to correct its environmental deficiencies, we have here either a volume which is not built and rigid until environmental power is applied to it, or a manufactured environment (conditioned air) and a bag to put it in. Either way, this might be claimed as a more subversive proposition than simply doing without built enclosure altogether, as in the case of the drive-in movie house, and by any standards it is a development alongside which most of the purely architectural revolutions of recent years must appear rather trifling, however hard won they may be.

Not that the ability to make inflatable architecture has been any less hard-won. The successful support of an air-structure involves not only the technologies required to fabricate the membrane and the air-handling equipment, but also considerable resources of knowledge and practical skill in the control of the air to be handled. Contrary to popular commonsense the pressures involved in air-supported structures are low, though the volumes of air to be supplied are large. The air-handling requirements are, in many ways, more akin to those of normal ventilation than to the high pressure air technology that most of us have encountered in mechanical experiences such as inflating motor-tyres or painting with a spray gun. But one has only to inhabit an air structure for a little while, and see and hear it accepting or resisting minute variations in pressure due to sun-heat, breeze, internal heating, door-flaps left an inch too far open or pulled down a half-inch too tight, to realise what delicate knowledge is involved in the management of this kind of environment.

Such knowledge, and the skill to apply it, has been accumulating since the turn of the century—indeed it is fundamental to the increasing use of environmental power, which must be subject to control precise enough for the fulfilment of the task to which it is directed. Unregulated power would have done as little to improve the conditions of men as unassisted structure. The achievement of such control requires a double knowledge—of the behaviour of the powered equipment being deployed, and the behaviour of the environment to which it is being applied. And thirdly, that these two sets of characteristics are mutually modifying. If there is a single historical turning point in the history of our practical understanding of this mutually modifying relationship, then the date that seems most proper is 1907. In that year, for a plant at the Huguet Silk Mills in Wayland, NY, Willis Carrier offered his first performance guarantee; that is to say, instead of treating the air-conditioning plant as another branch of structure, and offering to guarantee the quality of materials and workmanship, he faced the fact that what his clients were asking from him was to deliver reliably a certain kind of atmosphere, and offered to guarantee the quality of the environment instead. To do this he had to know not only the ability of his plant to handle air, and the environmental hazards promoted by the factory's machinery and work-force, but also, for the first time, the heat-gain due to the effect of the summer sun on the building's structure.

Though information sufficiently precise to support exact calculation proved difficult to acquire (throughout these early years of the art, one of Carrier's main labours was in producing sets of standard tables for all kinds of atmospheric calculations) he could, at the end of his efforts make statements as precise as

We guarantee the apparatus we propose to furnish you to be capable of heating your mill to a temperature of 70°F when outside temperature is not lower than 10°F below zero.

We also guarantee you that by means of an adjustable automatic control it will enable you to vary the humidity with varying temperatures and enable you to get any humidity up to 85% with 70°F in the mill in winter.

In summer-time we guarantee that you will be able to obtain 75% humidity in the mill without increasing the temperature above the outside temperature. Or that you may be able to get 85% in the mill with an increase in temperature of approximately 5°F above outside temperature.[3]

Though this does not promise absolute control (any humidity at any temperature) it makes a precise promise of sufficient control for the circumstances involved, and states most of the critical tolerances involved. Absolute control is rarely needed where the tolerances are known, and in an increasing number of situations in the period covered by this book, the critical tolerances are, quite simply, what human beings will tolerate. Whatever the part originally played by industrial installations like the Huguet Mills, the generalisation of improved environmental control has meant that the ultimate test of performance has been the subjective response of human individuals. If ventilating science began under the sign of the human nose, the ultimate end of total environmental control must be found under the sign of the whole man. But the whole man is not an ideal man, nor an average man, nor a man in any other way fixed and standardised. The objective of the growing battery of environmental sciences that have come to flourish in the last three decades is less to fix gratuitous standards for world-wide enforcement, like Le Corbusier's eighteen degrees centigrade, than to find what are areas of tolerable variation, how those variables are related to each other, and to the even more variable being they are intended to support.[4]

The environmental needs of the whole living man are variable in sickness and in health, youth and age, education and culture, physical and social circumstance. When British troops in Aden were lately accused of subtly torturing Arab detainees under interrogation by 'deliberately running the air-conditioning at "full cool"' it may well have been the case that the setting of the air-conditioner dial at 'full cool' was deliberate, and that the

[3] quoted in full in Ingels, op. cit., pp. 31–32.

[4] the history of physiological environmental studies—of human responses to heat, light and sound—remains to be written. The urgency of the need to get it written while the living memories of its pioneers are still available, is recognised by some of the pioneers themselves (like R. G. Hopkinson) but it will prove a formidable task.

Arabs, as result, felt subtly tortured, but the motives of the British troops may have been simply to make themselves feel comfortable, without possessing the necessary cultural and environmental insight to realise what this might do to persons raised in the local culture and climate. And the same British, or their close cousins, will complain of the 'ridiculous way' that Americans run their air-conditioning so cold that one has to remove clothes on leaving the building, without having the cultural and environmental insight to realise that only thus is it possible to wear, indoors, the mink stoles, etc., which are accepted badges of social rank in Texas and Southern California.

The recognition that there are no absolute environmental standards for human beings has required the environmental sciences to develop methods of assessing performance and needs that depend upon attempts to quantify subjective responses without doing injury to their human validity, to allow for the interaction of what is being assessed with other elements in the environment that are not under study, and to allow for variability in time through fatigue on the one hand, or conscious and unconscious accommodation on the other—faced with a glare of excessive light, one may reduce the amount of illumination, put on dark glasses, screw up the eyes or leave it to the contraction of the iris to compensate. Each of these may be the correct line of action, according to circumstance, and particularly as a function of the length of time to which one is exposed to the glare, for all tolerances seem to be greater where the extreme conditions occur only as occasional peaks in a flow of variables.

This combination of circumstance is fortunate for the shelter industry, since it means that, over time, the spread of tolerances is effectively wider than totally and minutely quantifiable laboratory tests might suggest. Given variation in time, the human body adapts itself to short term changes; the environmental control system does not have to make instant adaptation to every degree of temperature change in atmosphere or occupant, does not have to anticipate the effects of boiling a kettle or opening the fridge door. In many strictly lethal circumstances the time taken to get up from a chair, walk to a window and open it, is not a life and death consideration, and for less acute situations of vitiation or risk it may not be fatal to wait for someone else to become aware of the problem and open the window for you. It is possible that in the high risk conditions of hard vacuum the instant responses and omnicompetence of a space-capsule's life support system are absolutely necessary, and one knows what real life dramas the telemetering of an uncontrolled rise in cabin temperature can occasion, but here on Earth it will often prove that drawing a blind over a window, or actuating some other equally simple control, is all that is required. In the right circumstances, a truly sophisticated approach to the man/environment system may involve no complex mechanisms at all.

As has happened before (though not often enough) growing subtlety in our knowledge, and greater cunning in its application,

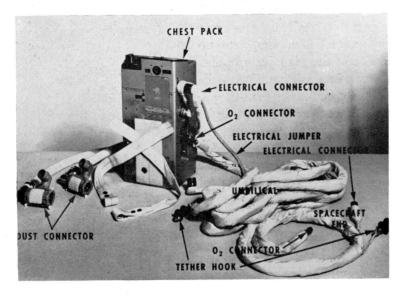

Space-walker's life support system, 1966.

Space suits, with air-conditioners for use on ground, 1965.

has made it possible to extract renewed increases in performance from time-honoured methods whose potentials might have seemed exhausted. Sooner or later the accumulated knowledge of man/ environment relationships that had been derived from the

St. George's School, Wallasey, Cheshire, 1961, by Emslie Morgan; the solar wall.

(necessarily experimental) application of new technologies becomes sufficient to facilitate a reassessment of traditional methods, and to suggest some imaginative re-deployments of the potentials thereby revealed. Those who have the knowledge are rarely (as some of them will admit) also equipped with the necessary imagination. Of the example about to be discussed, it has been said that any panel of accredited environmental experts to whom it might have been submitted would have found themselves bound to dismiss it as impracticable. The revenges of time are sweet, however, and established experts are reckoned to have spent more time and energy in trying to find out how it works than was ever lavished on it by its original designer.

The building in question is the second block of St George's County Secondary School in Wallasey (Cheshire, England). Completed in 1961, it belongs to that same generation of experimental environmental essays that were discussed in the previous chapter, but unlike them it has not enjoyed a world-wide press, doubtless because of the small fame of its designer, Emslie Morgan, principal assistant to the Borough Architect of Wallasey. Though he now has a secure reputation because the building has become something of a legend or *cause célèbre* among British environmentalists, he died before that fame was established, leaving no documents that can now be traced to record his thoughts and methods. The double lack, of both autograph documents and of any intelligent interest on the part of architectural publications when Morgan was alive, means that the present study can derive only from inspection of the structure as it stands and as it functions—and such inspection is becoming increasingly frequent.[5]

'Structure' is the word to emphasise, because what Emslie Morgan has offered in St George's School is an imaginative reappraisal of one of the oldest environmental controls known to man, massive structure functioning to conserve heat, plus an attempt at improved exploitation of the oldest and ultimate source of all environmental power, the sun. The structure is almost ludicrously heavy by the standards now current in British school building—nine-inch brick walls, seven-inch concrete roof, all wrapped in five inches of external foamed polystyrene insulation, plus further layers of cladding for various purposes. In plan, the block is long and narrow, with a slight bend at one point, and lies almost due east and west. The accommodation provided consists of classrooms and science laboratories for most of the length of the block, but the part beyond the bend contains a gymnasium and its ancillaries. The single-pitch roof stands barely high enough on the north face of the block to accommodate two normal storeys but on the south side it pitches up to over forty feet, thus providing a vast area of glass to the sun.

In the designer's mind, this 'solar wall' was undoubtedly the key to the functioning of the whole building, and has also been the aspect that has caught the fancy of the public. It consists of two skins of glass, separated by a space of 24 inches, the outer skin being clear, the inner one consisting almost entirely of obscured glass, to shed a diffused light into the teaching areas. Some of the inner skin is of clear glass, however, and at those points it is backed by opaque panels, painted black on one side, polished aluminium

[5] the Building Climatology Research Unit, Department of Building Science, University of Liverpool, has maintained the most constant watch on the school, and much of the information given here has been taken from, or confirmed by their Preliminary Report, in *Journal of the IHVE*, January 1960, pp 325 ff.

on the other, which are reversible according to the season, and are intended to provide a degree of thermal control by absorption/reflection of solar heat. Similarly, there are areas of the inner skin, in the assembly hall and gymnasium, that have been replaced by black-painted masonry, thermal performance being controlled by white wooden shutters that can be hung over them to reduce the absorption of solar heat.

It will be noticed that Morgan's use of glass avoids the traditional function of glazing—to be transparent to sight. There are, in fact, panes of clear glass in the hinged ventilation-windows that occur at intervals on both storeys of the façade, but they provide only scanty outward views. For this, and a tendency to overall glare from the glazed side of the rooms, the visual environment of the school has been subjected to some criticism. But about its thermal environment there seems to be no surviving doubt, now that its emergency hot-water heating system has been removed, unused, after the school had survived almost the worst winter in living memory (1962–3).

The heat so efficiently stored and managed by the massive structure has three main sources: the solar wall, the electric lighting, and the inhabitants. Of these, the solar wall may prove to be the least productive for most of the year, and the weak point in the school's armour of insulation in the cold of winter. The next most important source of heat is commonly taken to be the lights, which are switched on early to preheat the school before the pupils arrive, and some conservatively minded engineers have therefore described it as an electrically heated building. But the greatest source of heat is, in fact, the inhabitants themselves who, in a normally occupied class-room, provide about half the winter heat-input per hour. Even if it is the total management of the heat balance which is important here, the attempt to use the waste heat from the lights at a date well before the commercial availability of systems like Barber-Coleman Daybrite (which use heat-of-light to warm input air at the point of delivery) is worth a note in any

St George's School; close-up of solar wall.

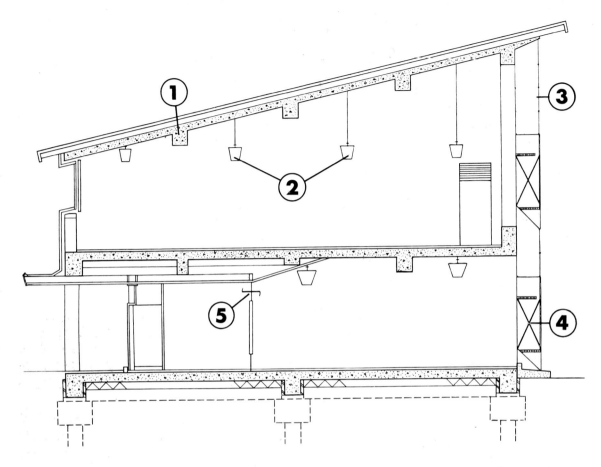

history of environment.

Nevertheless, it is the total view of the thermal environment of the complete man/structure/lighting/ventilating system that is impressive, as well as the simplicity of the methods for its control: a time-switch for the lighting's contribution to the diurnal heat balance, reversible panels for seasonal variations, and a card of instructions for each classroom on how the ventilation should be adjusted (by opening or closing the windows) to deal with short-term increases or drops of temperature.

St George's School; diagrammatic section of environmental provisions.

1. Insulated roof structure
2. Light fittings
3. Double skin solar wall
4. Adjustable ventilating windows
5. Ventilating windows at rear of classroom

One could object that this is too irregular and fortunate a case for any useful lessons to be learned from it; irregular in that it seems to work well but at variance with the designer's intentions for how it should work (as in the case of the solar wall), and fortunate in that it seems to enjoy both a site that is admirably suited to the proposition, and a local climate marginally more helpful to its working than many others might be, even in the same part of England. There can be no doubt that it is a special solution to a special problem, and less than perfect at that—difficulties with overheating on a few days of strong sun and no wind in high summer suggest that it needs a mild breezy climate even more than the direct incidence of sunlight for which Morgan designed it. But where is the building that does not have a few days of environmental difficulties in the year—by the going standards of environmental judgement, St George's School has proved itself as much of a success as any other building discussed in this book, and better than most.

Its successful performance guarantees its right to be discussed here, no more; the reason for discussing it is less that it works than because it works through the application of the ultimate form of environmental, and all other, power—knowledge. Even if Morgan were to prove mistaken in details, the overall proposition that he made presupposes knowledge of the total system so complete that one can judge what to omit—the heating system was never more than a hedge against unforseeable failure to function; it was never meant to be used and never was used. The professional courage to attempt such a radical reassessment of methods of environmental management can only come when quantifiable technological knowledge, derived from experience and controlled experiment, has acquired the same sort of completeness and authority as the accumulated rules of thumb by which vernacular cultures manage their environments.

After almost a century of such conscious, and consciously controlled, mechanisation of our environmental methods, we have a right to look for this kind of self-confidence on the part of architects, the self-confidence to reject the obvious mechanical solution because they know a better way—and not for the usual reason that they don't know enough about the mechanical methods available to choose the right one, or that they can't find one that fits in with their prefigured ideas of how the architecture should look. And architects will certainly need this kind of self-confidence if they are to make sense of the range of choice in environmental method now open to them.

To epitomise that range, let us simply review the examples cited in this chapter, all of about 1960 vintage, except the drive-in movie house which is a relative antique. The list covers:

Las Vegas; environment defined in light without visible structure of any consequence.

Drive-in Movie House; rally of mobile environmental structures in a space defined by light and sound.

AEC mobile theatre; space enclosed by membrane supported on a cushion of air.

Space capsule; rigid structure containing entirely and continuously manufactured life-support environment.

St George's School; massive structure conserving environmental output of the contained activities.

The extremes of this range, as represented here, are the AEC theatre, only one step removed from the pure application of power without any enclosure at all, and St George's School, only one step away from pure structure without any added power at all. Both extremes are demonstrably within the range that architects can professionally encompass—and this is not a question of visionary proposals about what architects ought to do in the future; this chapter shows that some have already done it. Nor may either of these extremes be dismissed as merely unique solutions to special problems, since all normal buildings are unique solutions to specific problems, and will remain so as long as buildings remain fixed to the ground in one place, which most of them will for a long time to come. Given this fixed location, every building will be exposed to an external micro-climate exhibiting some unique features, so that every building will be, to a greater or lesser degree, an unique environmental control system.

Of course, given lesser degrees of uniqueness and the human tolerances discussed above, it is possible to generalise the environmental requirements of quite large territories on occasions, and produce a building type that works pretty well all over that territory. Most of the vernacular houses of the past show this kind of generalised adaptation to a fairly well defined area—the Cotswolds, Western Norway, Central Japan, New Orleans. But these generalised adaptations are sometimes achieved only at the cost of human and social inconvenience—for instance, the heating arrangements in Japanese houses in very hard weather can immobilise the inhabitants in bulky clothing around a minute sunken pit containing an exiguous charcoal fire. And again, these generalised adaptations may still be insufficient to deal with particular adverse situations that arise within their home territory. This is especially true where the local type is applied too rigidly, for reasons of ancestral custom, status seeking or commercial inertia.

A classic case is the type of terrace housing evolved in the nineteenth century in Sydney, Australia. With their fronts protected against overhead sun by a projecting first-floor balcony and a roof pulled well down over the first floor windows, and against raking sun by the projecting party-walls that support both balcony and roof, they could hardly be bettered in that city's climate of extremes. But only so long as the fronts look North toward the midday sun. For the Sydney vernacular never evolved a matching solution for the rear elevations, and where these are the sunward side, the back rooms and tiny walled yards can become a kind of solar oven. So patent was this environmental failing that it contributed significantly to the fall from favour of the terraces, and now that they are regaining esteem because of their compact

Terrace housing, Sydney, Australia; structural sun-shading of street fronts.

urbanity, architects are having to find solutions to the problem of their hot backsides.

These solutions have so far shown considerable variation, and this is proper. Conscious architecture, as distinguished from vernacular building, should be able to reason out the unique solutions to specific problems. We should be sufficiently at ease with our basic kit of mechanical parts (as much at ease as Frank Lloyd Wright in 1910) to reverse the reflexes conditioned into us by the Masters of the Twenties and stop overvaluing such concepts as the norm, the standard, the *maison-type*; the more so since we now dispose of sufficient technology to make any old standard, norm or type habitable anywhere in the world. The glass skyscraper can be made habitable in the tropics, the ranch-style split level can be made habitable anywhere in the US. But this does not alter the fact that the California cottage remains demonstrably habitable as an alternative in California, and the Prairie House remains a desirably habitable alternative in Chicago. The existence of such strangers to current *maison-type* formats gives the freedom of a wider choice of environmental methods.

However, it is more commonly the mechanical alternative, not the structural, that gives freedom of choice. The structural alternative is normally the ancestral and restrictive vernacular which mechanism helps us to modify or replace with a solution better adapted to need, less restricting of function. Thus, the rules of orientation and plan-organisation for breeze, sectional organisation for cross ventilation and cooling, that apply to structural solutions in hot, humid climates, can become a tyranny that makes the sealed and necessarily mechanised envelope of a glass slab office tower look an extremely attractive solution, and in hot, dry desert climates its ability to exclude windborne dust from human activities that need to be kept clean, can make its attractions almost equally compelling. The present generation of experts on tropical architecture, conditioned by the experiences of dying colonialist régimes, seem to regard the glass skyscrapers that have appeared in developing countries as mere status symbols, the architectural equivalents of 'ignorant Fuzzy-wuzzy chiefs in top hats.' They may well be succeeded by a generation of experts on architecture in the temperate zones who wish that our Western civilization had been capable of making as bold a break with its ancestral vernaculars as the Africans have been.

Our ancestral vernaculars and our status-symbolising top-hats are the same thing—architecture as it is recorded in the history books. That tradition has one outstanding advantage over its newer rivals among environmental management techniques in that it disposes of a repertoire of symbolic forms—wall, roof, arch, column, vault—that can still bestow cultural status and power. But now that the techniques of unassisted structure have ceased to be the unique and inevitable solution to environmental problems, the unique force of those symbols has begun to wane. Hence the avidity with which Modernists, from Le Corbusier to the fantasists and visionaries of the nineteen-sixties, have stolen forms from other technologies—and hence too the inevitable disappointments when those forms proved neither to guarantee nor even indicate significant environmental and functional improvements over what the older structural technology afforded, because this was merely that older technology dressed up in borrowed clothes. But some, perhaps most, of the buildings discussed in this book show architects evolving, or beginning to evolve, forms which are not the borrowed finery of far-out technology, but forms proper to the environmental proposition being made, whether that proposition is as mechanically advanced as Lundy's inflatable pavilion, or as conservative, in the very best sense of the word, as Morgan's school. Only when such proper forms are commonly at hand will the architecture of the well-tempered environment become as convincing as the millennial architecture of the past.

Readings in environmental technology

As far as possible, all works consulted and sources of information have been acknowledged in the text or its footnotes, and any reader who wishes to pursue the topic further (and I hope that many will) should go to these sources. For those who simply wish to reinforce their understanding with background reading, the situation is less fortunate, because of the dearth of general works on the subject, about which complaint is made in chapter 1. However, the following works may prove helpful:

A Short History of Technology, by Derry and Williams, Oxford, 1960; especially chapters 14, 17 and 22, which give some account of water-supply, drainage, coal gas and electricity.

Home Fires Burning, by Lawrence Wright, London, 1964, which (together with his earlier *Clean and Decent*) gives an intelligent popularising over-view of aspects of environmental history. Nor can one ignore

Mechanisation Takes Command, by Sigfried Giedion, London and Cambridge, Mass., 1950, which—in spite of such spectacular shortcomings as a total failure to attack the history of electric lighting—still contains a mass of useful if ill-ordered information.

And, finally, a work intended for a general readership but now hard to find,

Willis Carrier, Father of Air Conditioning, by Margaret Ingels, Garden City, 1952, which contains an invaluable tabulated chronology of inventions and developments in ventilation and refrigeration from the Renaissance to 1950, to which the present study is deeply indebted.

Photo credits

Glasgow School of Art Library 84, 85

Henry-Russell Hitchcock 87, 88, 106, 110, 114–115

Esther McCoy (*ph* Marvin Rand) 102

Ulrich Conrads and Hans G. Sperlich 131

Walter Gropius 135, 137

A. W. Bruna & Zoon 138, 139, 140

R.I.B.A. Library 144

Robert Browning (*ph*) 157

Glaces et Verres (Et. Saint-Gobain) 161

Michael Carapetian/John Gallagher 165–167

Kenneth Frampton/A A Journal 164

Hedrich-Blessing (*ph*) 188 top, 211

Amana Refrigeration Co. Inc. 188 foot

Cervin Robinson (*ph*) 191

Baltazar Korab (*ph*) 192–193

Prof Julius Posener 201 top, 202, 203

David Gebhard (*ph*)/Schindler files 207

Richard Neutra 208

R.I.B.A. (*ph* Steiner and Nyholm) 210, 212

Dept. of Education and Science (*ph* Rondal Partridge) 217 top

Ezra Stoller 221, 224 foot, 236

Unations 225

J. E. Drew (*ph*) 226

C. F. Murphy Associates (*ph* Hedrich-Blessing) 229

Philip Johnson (*ph* Ezra Stoller) 230, 232

Alison & Peter Smithson (*ph* Nigel Henderson) 234, 254 foot

Architectural Press (*ph* H. E. Meyer) 66; (*ph* Sam Lambert) 258; (*ph* Stewart Bale) 281; (*ph* Cervin Robinson) 235

Peter Carter (*ph*) 238

Studio Zanuso 240, 241, 243 foot

Oscar Savio (*ph*) 245 top left, 246, 247

Cervin Robinson (*ph*) 248, 250

State University Construction Fund, Albany, N Y 253

Collection of the Museum of Modern Art, New York 256

Greater London Council 259, 262, 263

Curteicolor 271

Victor A. Lundy 272, 273

National Aeronautics and Space Administration 278–279

Wallasey Public Libraries 280

Max Dupain Associates (*ph*) 287

Index

Acoustic tile 195, 216
Adobe 24
Air-conditioners (domestic) 184–194
Air-conditioning 54, 55, 82–83, 102, 159, 162–163, 171, 194, 195, 209–213, 220, 222, 224, 242, 244–246, 257–262
Albini, Franco (and Helg) 239, 242–247
Alldis, Owen F. 58
Aluminaire House, Long Island 168–170
Amana air-conditioner 188
Anemostat (diffuser) 218
Archigram 256–257, 265
Argand lamp 55
Aristotle 29
Arizona Biltmore Hotel, Phoenix 197–199
Atkinson, Fello 244
Atomic Energy Commission Pavilion 270–273, 285

Bailey, George R. 182–183, 195
Baker House, Wilmette 106–111
Baldwin, William J. 39–40, 41
Balloon frame 101
Barber-Coleman (Daybrite system) 282
Bauhaus 123–124, 134–137, 139–142
Beecher, Catherine 96–100
Behrens House, Darmstadt 111–113
Behrens, Peter 86, 111–112, 253
Bell, Dr Louis 67–68
Bioscoop Vreeburg, Utrecht 139
Bird, Walter 270–274
Blanc, Charles 145
Botanical Gardens, Dahlem 126
Brandt, Marianne 139
Brawne, Michael 124
Le Braz, J. 162
Breuer, Marcel 122–123
Brise-soleil 156–158
Broderick, Cuthbert 176
Brooks, Morgan 69
Bullock's-Wilshire Store, Los Angeles 198
Bunshaft, Gordon 226
Burchard, John Ely (and Bush-Brown) 72 (note)

Burgess Acousti-vent (ceiling) 213–215
Burnham and Root 58

California architecture 93–95, 102–104, 198–201, 202, 204–208, 288
Camp fires 18, 20, 55
Carrier, Willis Havilland 26, 30, 46, 55, 82, 159, 162, 171–174, 180, 183–187, 222, 245
Carson, Arthur 185–186, 187
Chalk, Warren 256
Chareau, Pierre (and Bijvoet) 163
Cherne, Realto (and Chester Nelson) 180
Choisy, Auguste 122
Churches 33–34
Cité de Refuge, Paris 153, 155–158
Claude neon tubes 181
Coal gas 26, 32, 34, 55, 57, 58
Conduit Weathermaster System 180, 186, 222, 226, 228
Congrès Internationale d'Architecture Moderne 143
Conklin, Groff 100
Conservative Wall, Chatsworth 23
Continental Center, Chicago 228, 229
Villa Cook Paris 148–150
Cool air tank 117
Cooling 54
Cornell University Laboratories, Ithaca N Y 253–255
Cramer, Stuart W. 55, 82
Crompton, Dennis 256
Curtain wall 220

Dalsace House (Maison de Verre), Paris 163–168
Davidson, J. R. 199, 200–202, 204
Davidson, Samuel Cleland 14, 75, 81–82
Death-rate (industrial workers) 31
Desagulier, J. T. 51
Deutscher Werkbund 128, 130, 152
District heating 46–47
Diver, M. L. 178
Domestic lighting (consumption of) 55, 62
Doremus, Prof Ogden 174
Drake University Laboratories (project) 220, 222

Drexler, Arthur 252
Drive-in cinema 268, 285
Drysdale, Dr J. J. 35
Dymaxion House 96

Edison, Thomas Alva 26, 46, 60–64, 163, 141, 181
Ehrenkrantz, Ezra 216–217
Electric light fittings 66–69, 119–120, 136–141, 147–151,
206–208, 211–212, 234, 235
Electric lighting 43–44, 58, 70
Electric power 25, 53, 65 (domestic equipment)
Elliott, L. W. 219–221
Employment Exchange, Dessau 134–135, 140
Engelback, Norman 256
Equitable Building (Chicago) 228
L'Esprit Nouveau 145–150
Eugenie Lane Houses (Chicago) 192–194

Farnsworth House (Fox River Ill.) 228, 231
Feldman, A. M. 54, 195–196, 219
Finsterlin, Hermann 130
Fitch, James Marston 14, 24, 96–98, 187
Fluorescent lighting 181–183, 195, 216
Flügge, Richard 143
Forms (revealed by light) 69–71
Frampton, Kenneth 168
Franklin stove 25, 48, 97
Frantzen, Ulrich 253–255
Free Trade Hall, Manchester 176
Fuller, R. Buckminster 96, 265–266
Furniture Industry Building (project) 255–256
Futurism 124–125

Mrs T. H. Gale House, Oak Park Ill. 114–115
Gamble House, Pasadena 102–104
Garnier, Tony 86
Gas lighting 55–57
Gas mantle 57
General Hospital, Birmingham 76
General Motors Technical Center, Warren, Mich. 219–
221
Giedion, Sigfried 13, 14–16, 115
Gill, Irving 93, 94, 102
Glasarchitektur 125–129
Glass architecture 125–128, 151–156, 160–162
Glass Pavilion, Cologne 130–132
Gowans, Alan 94–95
Graumann's Metropolitan Theater, Los Angeles 177
Greene and Greene 102–104

Gropius, Walter 84, 86, 134–137
Grosse Schauspielhaus, Berlin 201
Guadet, Julien 268
Gulyas, Zoltan (and Szendroi) 254–255

Harrison, Wallace F. 224, 236–237
Harris, Thomas 'Victorian' 66–67
Harris Trust Building, Chicago 209
Hartog Study, Maarssen 136
Hayward, Dr John 35–38
Heating 38, 40, 45–52 (and see also: District heating,
Hot air heating, Hot water heating)
Heat-of-light 282–283
Henman and Cooper 75–76
Herpich Store, Berlin 201–202, 204
Herron, Ron 256
Hitchcock, Henry-Russell 93
Holborn Viaduct, London (electrical supply) 64
Holly, Birdsill 46
Holmes, Sherlock 31
Hopkinson, Prof R. G. 276 (note)
Horeau, Hector 143
Hot air heating 48–52, 114
Hot water heating 45–46, 105–108, 117–118
House of Commons, London 51, 174
Howe, George (and William Lescaze) 209–213
Huguet Silk Mills, Wayland N Y 275–276
Humidity control 24–25, 54, 81–82, 275–276
Huyett, M. C. 30

Inflatable structures 270–274
Inglenook 47
Inland Steel Building, Chicago 209, 228
International Style 94, 127, 129, 208

Jacob, Prof Ernest 31, 32–33, 34, 43, 48, 175–176, 177
Johnson House, New Canaan 228–233
Johnson, Philip 124, 228–233
Johnson Wax Company Offices, Racine Wis. 197
Jordy, William 209
Juhl, Finn 235

Kahn and Jacobs 223
Kahn, Louis 12, 239, 246–259
Kannel, Theophilus van 74
Kensington (electrical supply) 64
Kiesler, Frederick 20
Kimball, Dwight 41, 43
Kips Bay Apartments, New York 191–192

Kocher and Frey 168
Kroeschell (refrigerating plant) 91
Kuhn and Loeb Bank, New York 54, 195–196

Labasque, Yves 150
Lafayette Park, Detroit 188–189, 191
Lanchester, Dr F. W. 270
Larkin Administration Building, Buffalo 12, 27, 86–92,
    175, 249
Las Vegas 128, 269–271, 285
Lavoisier, Antoine 41
L C C Architect's Department 256
L C C housing, Roehampton 237–238
Lea, Henry 75
Lebon, Philippe 26
Le Corbusier 16, 97, 143, 145–163, 168, 171, 221, 224,
    237–239, 249, 265, 276, 287
Ledoux, Claude Nicholas 249
Lever House, New York 183, 209, 226–228
Lewis, Samuel R. 184
Light control 128–132, 141–142, 277
Light weight construction 100, 107, 117, 170, 231
Loos, Adolf 93, 95
Lovell beach-house, Newport Beach 204–206
Lovell House (Health House), Los Angeles 204–208
Lumiline tubes 181
Lundy, Victor 270–274, 289
Lyle Corporation Offices, Newark N J 163, 180
Lyon, Gustave 156, 160

Mackintosh, Charles Rennie 84–86
Madison Square Theatre, New York 177
Manuel de l'habitation 147, 168
Marinetti, F. T. 124–125
Marquette Building, Chicago 182
Massive structure 21–23
Maybeck, Bernard 102
McArthur, Albert Chase 197–198
McQuay air conditioners 186
Mechanisation Takes Command 13, 14–15
Meier, Konrad 42, 72
Mellon Institute (steel decking) 213, 214
Mendelsohn, Eric 200, 201–203, 204
Metropolitan Opera, New York 176
Milam Building, San Antonio 178–179, 209
Montauk Block, Chicago 58
Moore discharge tubes 181
Morgan, Emslie 280–284, 289
Murdock, William 26

Mur neutralisant 156, 159–162
Murphy, C. F. (and Associates) 228, 229
Murray, Matthew 45

Netsch, Walter 219
Neutra, Richard 170, 199, 200, 201, 207–208
Norwegian traditional architecture 100
Notre Dame du Haut, Ronchamp 238–239

Octagon, Liverpool 35–39, 56
Olivetti factory, Merlo 239–243
Opera house, Vienna 176, 177
Oud, J. J. P. 93, 194
Ozenfant, Amédée 145

Pavilion plan (hospitals) 76–74, 83
Pavillon Suisse, Paris 153–154, 155
Paxton, Sir Joseph 23, 143, 144–145
Pearl Street, New York (electrical supply) 64
Pei, I. M. (and Associates) 191–192
Performance guarantee 275–276
Perret, Auguste 86
Peters, Jacques 198, 199
Pettenkofer, Max von 42
Pevsner, Nikolaus 84–86
Pharmaceutical Laboratories, Debreczen 254–255
Philadelphia Savings Fund Society Building, Philadelphia
    180, 209–213
Physiological studies 276–278
Plenum system (ventilation) 52, 72, 76 (William Key's),
    77, 84, 175, 259
Plug-in aesthetic 257
Poelzig, Hans 130, 201
Pollution 29
Prairie houses 104–121, 288
P S A L I 182

Queen Elizabeth Hall, London 238, 256–264

Radio City, New York 162, 181
Railway coaches 175, 185
Regenerative mode (of environmental control) 23
Reliance Building, Chicago 73
Revolving door 74
Richards Memorial Laboratories, Philadelphia 12, 239,
    246–255, 257
Rietveld, G. T. 136–139, 140–141, 204
Rinascente Store, Rome 239, 242–247, 257, 258
Rivet Grip Company, 213

Isobel Roberts House, River Forest, Ill. 112, 114
Robertson Q-deck 219
Robie House, Chicago 115–121, 197, 232
Van der Rohe, Mies 130, 188–189, 191, 228, 237
Ross Cottage, Delavan Lake 110–112
Row-houses, Utrecht 140–141
Royal Victoria Hospital, Belfast 27, 75–84, 86, 175, 240
Rudolf Mosse Offices, Berlin 201
Rumford fireplaces 25, 48

Saarinen, Eero (Saarinen, Swanson and Saarinen) 219–222
St George's school, Wallasey 280–284
Saint-Gobain tests 160–162
Salle Pleyel, Paris 156
Sardi's Restaurant, Hollywood 204
Villa Savoye, Poissy 150–151
Scheerbart, Paul 20, 28, 125–132, 142, 143, 231, 270
Schindler, Rudolph 170, 199, 200, 204–207
School of Art, Glasgow 84–86
Schools Construction System Development 216–217, 240
Villa Schwob, 158–159
Selective mode (of environmental control) 23, 24
Servant spaces 114, 249, 253
Sheffield University (project) 254, 255
Ship-building technology 266–267
Sirocco fans 32 (and see also: Davidson, Samuel Cleland)
Skidmore, Owings and Merrill 209, 219, 226–228, 237
Skyline Louverall ceiling 218
Skyscrapers (environmental problems) 72–74, 181
later, John 58–59
Smithson, Alison and Peter 234, 254–255
ow, William Gage 48–49, 51
ar wall 281–282
h-eastern U S (traditional architecture) 24
ace technology 268, 278–279
am, Mart 14, 200
eam Hall, Leeds 45
e Stijl 132
iftsgarden Palace, Trondheim 100
ock Exchange, New York 174
Stokesay Court, Salop 66–67
Studio houses, Paris 152
Sturtevant Company 50, 51, 52, 53
Sun control 121

Suspended ceiling 195–196, 213–219, 220
Swan, Joseph 58, 60

Taut, Bruno 130–134
Taut House, Berlin 132–134
Teale fireplaces 33
Technology (as a cultural problem) 122–123
Tents 18
Terrace housing, Sydney N S W 286–287
Tesla, Nikola 53
Theatres and cinemas (ventilation) 175–178
Thompson, Prof Elihu 67
Timber construction 25, 100–101, 107
Town Hall, Leeds 176
Tropical architecture 288
Turbinenfabrik, Berlin 253

Unité d'Habitation, Marseilles 237
UN Building, New York 183, 221–226, 234–237
Universal Pictures Building, New York 220, 223
Universum Cinema, Berlin 203

Valéry, Paul 266
Ventilating (fan) 51–54
Ventilation 24, 32–34, 40–43, 109, 120 (and see also: Ventilating (fan) and Plenum system)
Vernacular environmental controls 286–287
Vers une Architecture 147, 148, 163
Victoria Regia house, Chatsworth 144–148
Voysey, Charles F. A. 47

Ware, Isaac 268
Watt, James 45
Webb, Michael 255–256
Weese, Harry 192–193
Weissenhof exhibition, Stuttgart 152
Welsbach, Baron Auer von 14, 57
West End Cinema, London 14, 181
Willis, George 178
Wilson, Colin St John 249, 253
Dr Winter 145–146, 150
Winzer, F. A. 26
Wolfe, Tom 269
Wright, Frank Lloyd 12, 24, 47, 70, 86, 93–95, 104–121, 194, 197–200, 230, 288

Zanuso, Marco 239–243